# SpringerBriefs in Applied Sciences and Technology

SpringerBriefs present concise summaries of cutting-edge research and practical applications across a wide spectrum of fields. Featuring compact volumes of 50–125 pages, the series covers a range of content from professional to academic.

Typical publications can be:

- A timely report of state-of-the art methods
- An introduction to or a manual for the application of mathematical or computer techniques
- A bridge between new research results, as published in journal articles
- A snapshot of a hot or emerging topic
- An in-depth case study
- A presentation of core concepts that students must understand in order to make independent contributions

SpringerBriefs are characterized by fast, global electronic dissemination, standard publishing contracts, standardized manuscript preparation and formatting guidelines, and expedited production schedules.

On the one hand, **SpringerBriefs in Applied Sciences and Technology** are devoted to the publication of fundamentals and applications within the different classical engineering disciplines as well as in interdisciplinary fields that recently emerged between these areas. On the other hand, as the boundary separating fundamental research and applied technology is more and more dissolving, this series is particularly open to trans-disciplinary topics between fundamental science and engineering.

Indexed by EI-Compendex, SCOPUS and Springerlink.

More information about this series at http://www.springer.com/series/8884

Shuichi Fukuda

# Self Engineering

## Learning From Failures

Springer

Shuichi Fukuda
Systems Design and Management
Research Institute
Keio University
Yokohama, Kanagawa, Japan

ISSN 2191-530X ISSN 2191-5318 (electronic)
SpringerBriefs in Applied Sciences and Technology
ISBN 978-3-030-26724-7 ISBN 978-3-030-26725-4 (eBook)
https://doi.org/10.1007/978-3-030-26725-4



This Springer imprint is published by the registered company Springer Nature Switzerland AG
The registered company address is: Gewerbestrasse 11, 6330 Cham, Switzerland

# Preface

The core messages of this book are as follows.

(1) Let us live as humans. We are not machines. We perceive environments and situations, and we make decisions, based on our motivation. We decide what action we will take.
(2) Let us enjoy our life. But let us keep on growing the world of human species. We should try more to actualize ourselves. The efforts of each "Self" contribute to expanding human world and human culture.
(3) As environments and situations change frequently and extensively in an unpredictable manner, let us make the most of our "Instinct" endowed by Nature. This is the tool for us to survive and that is "Evolution." "Evolution" is nothing other than "Adaptation."
(4) We should establish "Self-Supporting Society." "Self-Supporting" is completely different from "Self-Sufficiency." "Self-Sufficiency" is defined with the current industry or engineering framework as its background. You choose to make your life "Self-sufficient," because you would like to, or you had to. But the world outside does not change.

"Self-Supporting" completely changes the whole industry or engineering framework. Current industry or engineering framework is based on the principle of "Work for others," in other words, "Division of Labor." But "Self-Supporting Society" is based on the principle of "Work for myself", You work for yourself, Although even if everybody works for himself or herself, the labor is appropriately shared. So, the human world develops and expands.

3D printing is discussed from the viewpoint of current industry or engineering framework to increase productivity. But it can be one of the tools to provide "Work for myself" opportunities to us.

To accelerate this shift, engineers need to develop technologies that enable us to use easily in our daily life. Certainly, we need chefs on some occasions. But in our everyday life, it is more important for us to know how we can cook the leftovers in the refrigerator and have nice dishes. This is a challenge and indeed provides us with great satisfaction, when we succeed.

This is the same idea as "Sharing Economy." Here, I would like to call it "Sharing Life." We do not need to stick to a single life. We can enjoy a multiple life. Engineering is making our dreams come true. It is said only humans can see the future. Then, why don't we enjoy the life of tomorrow now?

What makes this possible is "Instinct" endowed by Nature. "Instinct" is common to all humans. Personalization is fine, but we should devote more efforts to "Self" activities, which are inherent in human species.

Thus, this book is written for you to realize another engineering philosophy is needed to develop and establish World 2.0, which is a new world of tomorrow.

I hope you will enjoy reading this book.

I would like to thank many people who gave me new ideas and new perspectives through discussions.

They opened the door to the new world.

Finally, I would like to thank Mr. Anthony Doyle, Springer, London, and Ms. Megana Dinesh, SpringerNature, Chennai. Their encouragement and patience are truly appreciated.

Yokohama, Japan
June 2019

Shuichi Fukuda

# Contents

# Chapter 1
# Introduction

When you saw the title of this book, you might have wondered "Self Engineering? What is it? Is it another Do It Yourself?" The answer is Yes and No.

"Do it yourself (DIY)" is the word of the producer. Its primary purpose is for the producer to satisfy the growing needs of diversification and personalization of the customer. This is fundamentally a word from the producer to the customer. This symbolizes the current industrial framework where the producer and the customer are divided.

The word for "Self Engineering" is "Do it myself (DIM)." You are motivated internally to do the job and you, yourself, make a decision to do the job. It is deeply related to the highest need of humans, "Self-Actualization," which Abraham Maslow pointed out [1, 2]. The importance of "Self" is pointed out by Edward Deci and Robert Ryan in their Self-Determination Theory [3].

Why they emphasized its importance is because it is related to our need as a living creature. All living creatures are born to expand their world and to adapt to the changes. "Adaptability" is getting attention these days. But all living creatures are born to adapt to the changes of the environments and situations. That is "Evolution." "Evolution" is nothing other than "Adaptability." "Self" in this sense is our need as the human species and all living creatures are living on "Self."

Current engineering focuses too much on products. But as "Sharing Economy" demonstrates, we do not need so many cars. If we can share cars, we can enjoy driving cars that best fit the environment and the situation. When cars will become electric vehicles (EVs), we can enjoy assembling a car just like Lego. Then, car component companies can produce their components in mass, but they do not need such large factories as you see today. We can enjoy not only the product, but the process itself. This is another "Self-Actualization" or "DIM."

In other words, we can enjoy making our dreams come true. But engineers forgot why we wish to make our dreams come true. It is said that only humans can think about the future. It is one of the valuable resources endowed by Nature to explore and to expand the world of the human. Engineers, however, are enjoying their own

 
S. Fukuda, *Self Engineering*, SpringerBriefs in Applied Sciences and Technology,
https://doi.org/10.1007/978-3-030-26725-4_1

"Self-Actualization" and focus on developing higher technologies and pay most of their efforts.

Have you ever considered the electricity consumption needed for Watson and AlphaGo? They consume 100–200 kW. Humans, on the other hand, only consume about 20 W. Watson or AlphaGo is splendid. But if we consider the current trend of Robotic Process Automation (RPA), I believe we should re-examine whether we really can sustain such tools or keep on automating which consume tremendous amount of energy. This is because our current society is divided into two groups: the experts and their users.

If we can remove these walls between experts and between experts and users and change the society to "DIM," then everybody can enjoy their life his or her own way and still keep the energy consumption to the minimum. In short, our society will change from "Working for others" to "Working for myself."

"Self-Actualization" is given by Nature to explore and to expand the world of human species, so if we utilize this "Instinct," we could develop another society, which consumes the minimum energy, but still provides enjoyment and satisfaction to humans.

To achieve this goal of developing such a new world, we need to go ahead by trial and error. In such a world, humans will be a playing manager in Human–Machine Mixed Team. To make such a dream come true, we need to learn by failures.

This book describes how engineering is changing and what the future seems to hold for us.

And finally and most importantly, this book introduces a pattern-based approach to learning from failures.

## References

1. A.H. Maslow, A theory of human motivation. Psychol. Rev. **50**(4), 370–396
2. https://en.wikipedia.org/wiki/Maslow%27s_hierarchy_of_needs
3. https://en.wikipedia.org/wiki/Self-determination_theory

# Chapter 2
# Engineering Yesterday and Today

## 2.1 Late Nineteenth and Early Twentieth Century

Very late in the nineteenth century, a wide variety of new products emerged, which play important roles in today's society. For example, 1886 was regarded as the birth year of the modern car when German inventor Karl Benz patented Benz Patent-Motorwagen [1, 2] (Photo 2.1). And the Wright Brothers flight was in 1903 [3] (Photo 2.2).

The late nineteenth century was the age of inventions. Therefore, "Self" was important. Although the word "Self" was not popular, people placed importance on "Self."

But this period when "Self" was important did not last long. It is a big surprise that Ford Motor started to mass-produce cars in 1901 [4, 5] (Photo 2.3). And many inventions came to be produced in mass. Mass production calls for uniformity, and the "Self" came to be ignored. This trend continued until the late twentieth century.

Twentieth century may be called an age of conversation. Our world was small and closed with boundaries. So, we share the language and we can exchange our thoughts as symbols (Fig. 2.1).

There were changes in these days, too. But these changes were smooth, so that we could differentiate them mathematically, i.e., we could predict the future (Fig. 2.2). Therefore, engineers could foresee the operating conditions of machines and mass-produced products were easily accepted.

Machines were designed and produced on an individual product basis. So, we could give instructions to each machine from outside and machines responded. To describe it another way, we give instructions to machines as engineers designed. So, there is no basic difference between machines and human operators. Human operators' role was the same as that of machines. Humans were just another machine. Therefore, there was no room for "Self."



S. Fukuda, *Self Engineering*, SpringerBriefs in Applied Sciences and Technology,
https://doi.org/10.1007/978-3-030-26725-4_2

**Photo 2.1** Benz Patent-Motorwagen

**Photo 2.2** The Wright Bothers

**Photo 2.3** Ford motor T mass production

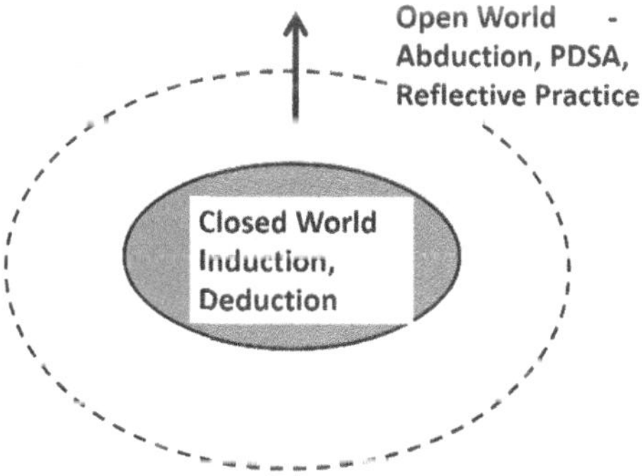


**Fig. 2.1** Closed world and open world

The word "Self" was used in a different meaning in these days. Such words as "Self"-operating machines, for example, were frequently used. But this does not mean machines came to be inventive. Completely the other way around. It just means that machines operate as engineers designed without any additional instructions from the operator. It only means that we can save time and effort to give instructions to machines every time.

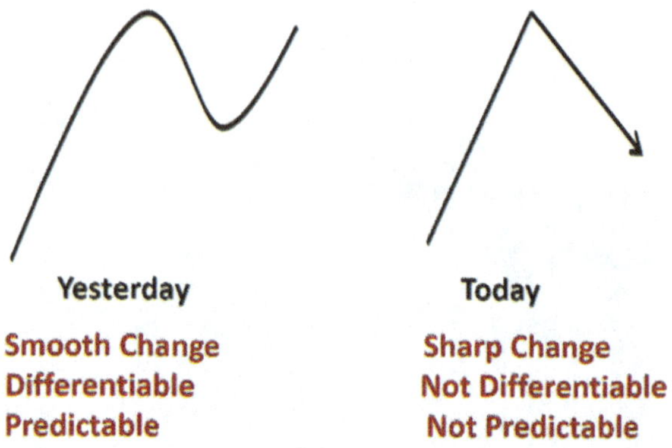


**Fig. 2.2** Changes of yesterday and today

## 2.2 Mid-Twentieth Century

The World War II ended in 1945. But facilities and equipment were broken or severely damaged. And food is difficult to secure. So, we had no other choice but to secure food, etc., on our own. In short, we need to practice what Theodore Roosevelt told us as the soul of exploration, "Do what you can, with what you have, where you are." We were practicing exploration.

These experiences served as a springboard to the exponential progress of engineering in the late twentieth century.

## 2.3 Late Twentieth Century

This ignorance of "Self" prevailed until mid-twentieth century. Mass production was the major engineering strategy these days. But the rapidly progressing engineering expanded our world, and our world became an open world without boundaries (Fig. 2.1).

This changed engineering. Our engineering until then was just like agriculture. We have our land and cultivate it in pursuit of a rich harvest. The goal is fixed, and we pay our attention and efforts to find a better way. In fact, agriculture is the first sector of economy and engineering developed as a second sector to develop it (Fig. 2.3).

But with our world quickly shifting to an open world, engineering shifted from agriculture to hunting. In hunting, the goal or the game is not fixed. We may come across animals which are stronger than us and we may be a prey, instead. And we need different ways to catch or kill different games (Fig. 2.4).

Thus, expansion of our world revived the importance of "Self." "Self" determines which game to catch or kill or to escape, depending on what animal "Self" comes across.

This change may be compared to railroad and voyage. In the case of a railroad, the destination is clear and what we have to do is to determine which line and which train

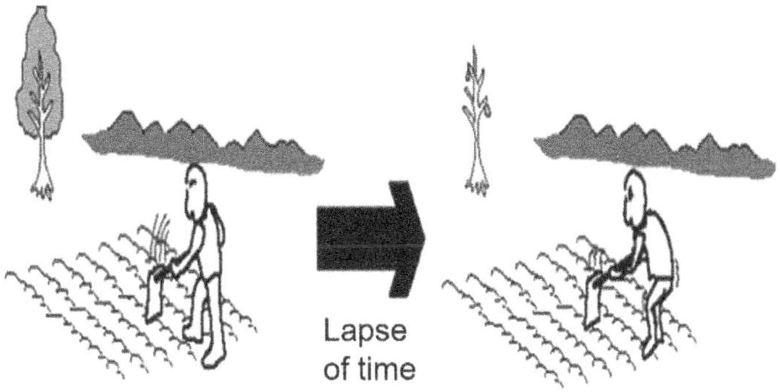


**Fig. 2.3** Agriculture

**Fig. 2.4** Hunting

to take. So, "Self" is not called for. Anybody can make a decision at the beginning of a journey.

But in the case of a voyage, the weather may be very good today, but tomorrow, hurricane might come. Then, we need to evacuate and set sail at some port nearby. In the worst case, we have to abandon the originally planned goal and head for another goal with the changing situations. Thus, in voyage, we have no other choice than to go by trial and error (Fig. 2.5).

Hunting and voyage need adaptability. As environments and situations vary from case to case, perception (situational awareness) and decision making become crucially important. Therefore, "Self" becomes very important. In fact, "Self" only knows what is happening right now.

Fig. 2.5 Railroad and voyage

## 2.4 Early Twenty-First Century

Such a shift quickly removed the walls between different areas. We must take note that such a shift is not diversification on the individual basis. Diversification and personalization are very often discussed together. But we should be careful that personal diversification and the world-level diversification are different. What is important for human species is this world-level diversification. Many different areas are now working together as a team. It is not a team of individual members. But **a team of teams**. How we respond to this problem of a team of teams is a big challenge.

Let us take a simple example of brand. Yesterday, brand was the quality assurance. You are assured to receive the same quality of products as you did. But today, such reproducibility lost importance. Adaptability becomes most important. How a company can produce products that adapt very well with the changing environments and situations becomes crucially important.

Basically, this is what "Self" is practicing. How human species can survive in the frequently and extensively changing environments and situations is what "Self" is pursuing. In other words, that is "Self" is pursuing how to evolve.

## References

1. https://en.wikipedia.org/wiki/Car
2. https://en.wikipedia.org/wiki/Benz_Patent-Motorwagen
3. https://airandspace.si.edu/exhibitions/wright-brothers/online/fly/1903/
4. https://www.history.com/this-day-in-history/fords-assembly-line-starts-rolling
5. https://silodrome.com/ford-model-t-assembly-line/

# Chapter 3
# Self

## 3.1 Human Needs

Maslow proposed a hierarchy of human needs [1] (Fig. 3.1).

At the bottom level, humans look for material satisfaction. They look for food, etc., to survive. But after these needs are satisfied, they go up the ladder. Their needs shift gradually from material to mental. The primary sector of the economy is agriculture, fishing, etc., as everybody knows. Engineering developed as the secondary sector to make the most of these natural resources. The tertiary sector is the production of services instead of end products [2].

At the top of the pyramid, humans look for "Self"-actualization. They would like to demonstrate how much capable they are. Therefore, to take mountain climbing for example, mountain climbers choose a difficult route and once they succeed, they challenge a more difficult route the next time. And the more difficult the route is, the more satisfaction they get. If they only want to reach the top of the mountain, they could ask a helicopter to take them there or they could take a much easier route. So, challenge is the core and mainspring of all human activities. And that is the same reason why we love games. We would like to demonstrate how capable we are. But we should remember we are born to act this way. This is what each "Self" is contributing to expand the community of the human species, although we are not aware of it.

It is interesting if we note that the year Maslow published this paper is 1943. Around this time, mass production spread widely, and inventive "Self" became ignored. Products were evaluated based on the functions, and reproducibility was important. Products were expected to work the same way, no matter how situations might change. Such reproducibility-central trend was prevalent because situations did not change appreciably, and changes were predictable.

It is interesting, too, that in 1944, Johan Huizinga published "Homo Ludens: A Study of the Play-Element in Culture" [3]. He pointed out the importance of "Play" for



S. Fukuda, *Self Engineering*, SpringerBriefs in Applied Sciences and Technology,
https://doi.org/10.1007/978-3-030-26725-4_3

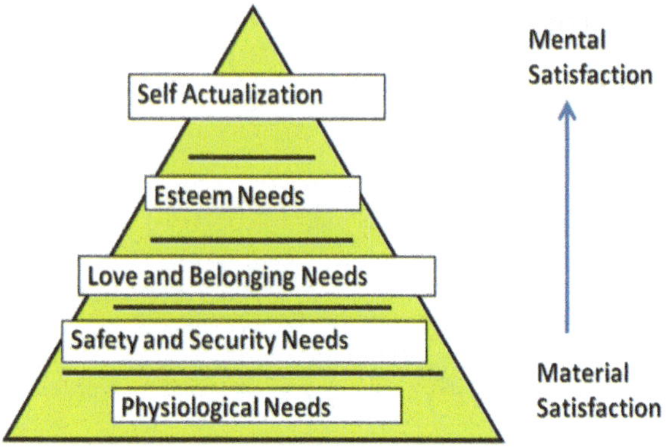


**Fig. 3.1** Maslow's hierarchy of human needs

developing "Culture." We must remember "Culture" and "Civilization" are different. "Civilizations" come and go, but "Culture" persists, because it is related to human species. Culture is the community of human species.

We play to expand our community. Look at babies and children. They play to explore the new world. They play to grow.

"Growth" is another form of "Evolution" or "Adaptability."

Ironically, this overwhelming mass production opened our eyes to the importance of "Self."

In 1985, Deci and Ryan proposed Self-Determination Theory [4, 5]. Even if the job is the same, we feel much happier, if we are motivated internally and make decisions ourselves to do the job, than we are told to do the same job from outside, no matter how much high the reward may be. They made clear the importance of intrinsic motivation and self-determination.

## 3.2 Personalization

The discussion above about "Self" should not be confused with personalization. The expansion of our world brought diversification, and people's requirements come to be personalized. And such marketing strategy as mass customization is proposed.

This is the discussion of the "personal" diversification. Our requirements vary from person to person by nature. The "Self" we are discussing here is so to speak what we share across individuals as human species. Personalization is a problem of individual variations and is not a problem of the whole human species.

Mass customization is a marketing strategy to sell mass products to customers with diversified requirements [6]. It is very much extrinsic. What we are discussing here is intrinsic "Self."

## 3.3 Creativity

Mass customization was an effort to make mass production and personalization compatible. But we must remember why Maslow emphasized the importance of "Self"-actualization and Deci and Ryan pointed out the importance of intrinsic motivation. All living things make efforts to expand the community of their species. Humans are no exception. All living creatures are born to expand their world. Thus, they are all explorers. That is why they are called living creatures. Creatures comes from the Latin "Creare (-Create)." Thus, they are creative to develop their species.

### *3.3.1 Creative Customers*

Kids are very creative. Even when they go down the same slide, they invent many different ways of sliding down (Fig. 3.2).

Youngsters also invent their own ways of enjoying wears. Jeans are used for working. So, they make holes in their jeans to share stories (Fig. 3.3).

**Fig. 3.2** Kids are very creative

**Fig. 3.3** Youngsters enjoy stories

### *3.3.2 Sharing Economy*

Sharing Economy is getting wide attention these days. This is another example of how creative customers are. They realized there are many ways of enjoying cars. We used to buy cars and drive them. Passenger cars are so to speak a status symbol. Customers' expectations did not change from customer to customer. But today customers realized their motivations vary from time to time. Sometimes they would like to drive a sports car, and other times they would like to enjoy driving with their families. They realized what they expect from cars vary widely from time to time and from situation to situation. Then, they can share a car and drive it when they need that kind of a car.

## 3.4 Modularization

Although the basic industry framework of mass production did not change after World War II, the facilities were destroyed during the war, so mass production changes in its strategy. Before the war, it was simply mass production. Products were designed and produced in mass. No ingenuity was needed. But after the war, to keep the framework of mass production with many factories and equipment severely damaged, new strategies emerged. One is modularization, and the other is model upgrading.

### 3.4.1 VW Factory in Resende, Brazil

It is said that modularization started when Volkswagen (VW) opened a new factory in Resende in Brazil in 1996. Local industries were very much afraid VW would destroy their industries, but VW took a very much different strategy. They rented the factory space to local industries and leave the production operation totally to local industries. Local industries formed a production team and what is very unique, VW asked them to sign a contract on the basis of final product sale.

Usually, such component suppliers sign contracts on component basis with the assemblers running a factory. But VW was very clever. If they leave everything to a team of local companies and contract on the final product sale base, then VW does not have to worry about how they can adapt to Brazilian customers. Local companies know very well what Brazilians want. So, VW sent only several Germans to run the factory. This approach was very much welcomed by Brazilians, and VW Resende factory was called "Dream Factory" [7, 8]. But from VW perspective, they can save a lot of money by introducing such an approach, and they can produce a car Brazilians want in much shorter time and with much less cost. So, it brought win-win relation.

Modularization was introduced in Resende factory to facilitate the team working of local industries. Later, modularization was changed to function-based, but initially it was how the production can be adapted to the local production customs and procedures. We must remember such modularization strategy plays an important role when we go abroad and tie up with local industries. Such win-win relation-focused approach should be utilized more in modularization. In fact, this way of modularization was born after VW's long experience and struggle abroad.

But such Resende style modularization did not spread much. Instead, function-based modularization spread very rapidly and widely. It is because if we introduce this strategy, we can produce parts in mass, and by combining them differently, we can easily cater to personalization. Mass customization is basically marketing strategy, but this modularization is producer-centric.

Another reason for such acceleration of modularization is industries come to design and produce products as a team. Yesterday, products were designed and produced on an individual or single kind basis. But today they are designed, produced, and operated as a team.

To describe it another way, our product development used to be a tree structured, but today how we form a network of a product team becomes important. In the case of individual product development, a tree structure is best, because there is only one output node and every node (member) are supposed to contribute to this goal.

But in the case of team products, its structure is basically a network. In a network, any node can be an output node, so that we can deal with the rapidly progressing diversification (Fig. 3.4).

Such part-centric modularization opened our eyes to a new world. We have been focusing our attention on the performance of final individual products, but modularization leads us to identify other values we have not been aware of.

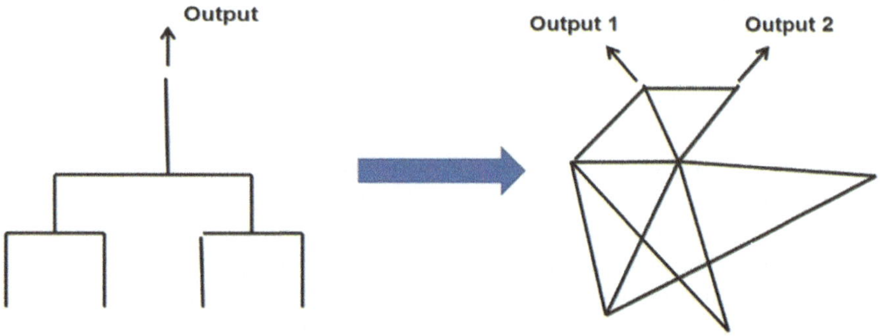


**Fig. 3.4** Individual to team product (tree to network)

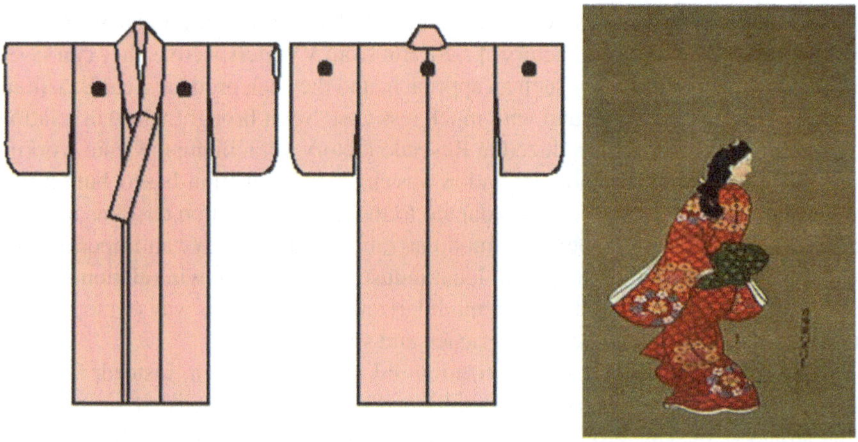

**Fig. 3.5** Kimono

### *3.4.2 Kimono*

Kimono is another old example of modularization. Western dresses have been designed to fit the body, but kimono (Fig. 3.5) is made of four main strips of fabric (modules) and we put it on. And by changing combinations, the same fabric can be re-tailored, and the grandchild wears the same kimono her grandmother wore.

### *3.4.3 Wedding Dress, Fashion Show*

Wedding dress is another good example of modularization. Every lady would like to wear a wedding dress just for herself. But not every lady is rich. So, they have to put up with the rental dresses. Rental dress shops are very well aware of the

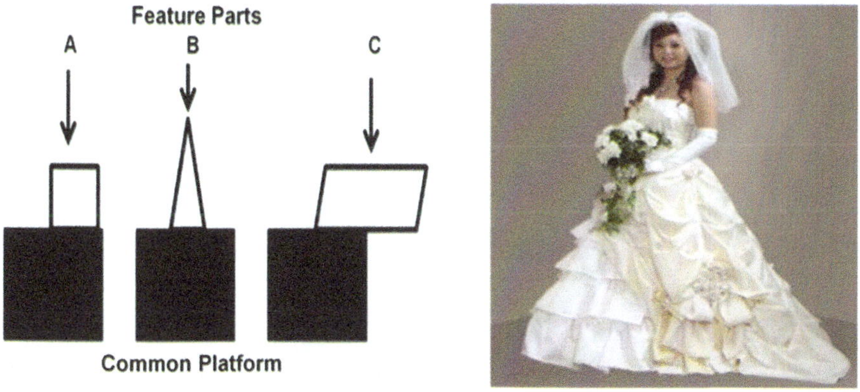


**Fig. 3.6** Wedding dress

bride's wishes. So, they hold a fashion show and observe where the participants pay attention to. They produce a wide variety of these feature parts. But the other parts which do not draw attention are identical. They constitute a common platform. We have to realize once again that modularization is not for making production easier or cheaper, but to satisfy the self-actualization need of humans (Fig. 3.6).

### *3.4.4 Accessary*

Accessary is the same idea. Everybody would like to express herself or himself best in the environment and in the situation.

### *3.4.5 Cars*

The word "Modularization" came popular with car industry. In fact, modularization is often discussed these days, but most of them are related to passenger cars. Trucks introduced modularization long time ago. But we should be careful. The modularization in the case of trucks is to facilitate the production, and it is not basically related to the self-actualization need. In the case of a truck, different bodies are needed for different users. So, chassis and bodies are produced in different factories, and they are assembled later (Fig. 3.7).

Modularization of a passenger car has been developed on the same idea as trucks. Modularization facilitates production. But these days, another kind of modularization is emerging. That is the same idea as accessary. Users can enjoy how they can combine the modules in different ways to best fit the situation (Fig. 3.8).

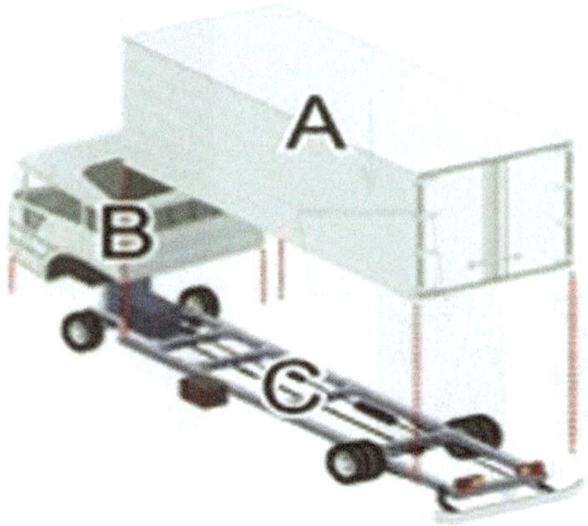


**Fig. 3.7** Modularization of a truck

**Fig. 3.8** Self-assemble car

## 3.5 Process Value: Its Increasing Importance

The framework of industry has long been linear. This is because the customer needs are at the bottom of Maslow's hierarchy of human needs, and they wanted material satisfaction. In other words, they wanted products and the products they needed were very clear. So, engineering developed to process the natural resources at first. And even when engineering developed further, the functions of the final products have been most important.

Products were expected to work the same way, no matter how the environment and situations may change. Thus, the goal was very clear, and problem solving was most important in these days. Further, material engineering was not so much developed, so engineers had no other choice than to produce their products with available choice of materials. Due to these backgrounds, industries introduced linear system (Fig. 3.9), and they managed their systems based on a tree structure (Fig. 3.4).

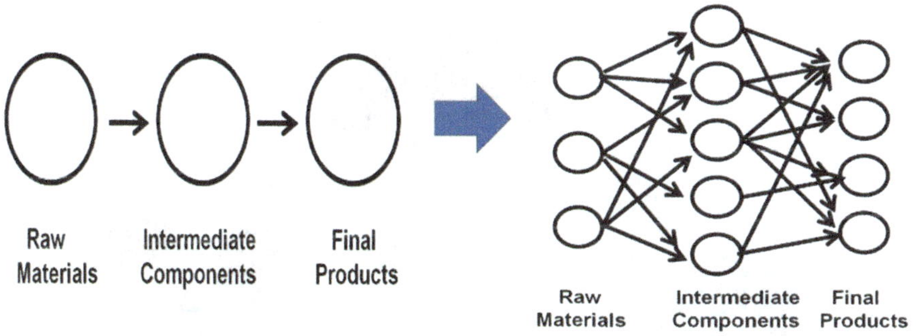


**Fig. 3.9** Change of industry framework: from linear system to parallel processing system

Although engineering developed rapidly, this basic linear system industry framework did not change until very recently. The efforts were paid to reduce time to market and reduce cost. For example, DARPA proposed DARPA's Initiative in Concurrent Engineering (DICE) in 1988. Its purpose is to reduce time to delivery by bringing downstage processes more upstream and process the jobs earlier.

But recently, material engineering developed so much that material engineers can produce almost any materials engineers want. So, the restriction of material choice has been removed almost completely. And engineering expanded our world enormously so that our life space expanded so much. Therefore, the products we need are quickly diversified. Modularization is one of the solutions to meet such diversifying needs. Thus, industry framework is quickly changing to Parallel Processing System (Fig. 3.9). The Parallel Processing System is nothing other than neural network.

This change of industry framework led us to the new world of value. The value yesterday was product value. How a product works best was the primary interest of our customers. But now customers find there is another joy, which is more exciting, and which satisfies their "Self" needs. Now, industry is ready to provide them with parts which customers can combine by each customer's way and to come up with what they want. Therefore, engineering changed from product-centric to process-focused.

Yesterday, engineers regarded modularization as a good tool to adapt to the diversification of customers' requirements. Their focus was on final products alone, and their view is from the manufacturing aspect. They did not consider what the customer really wants. This is the problem of "Self," not personalization.

### 3.5.1 *Lego*

It is very much interesting and surprising that Lego production started in such early times as 1947. Lego is very much different from the philosophy of industry until then. Lego just produced blocks and they sold them to customers. No products were sold. Just blocks. So, doing business with a considerably reduced number of factories was not a problem at all and further they achieved the best cost performance (Fig. 3.10).

**Fig. 3.10** Lego

Customers enjoyed very much. They enjoyed the process of combining different parts to produce what they have on their minds. This satisfied their creative "Self," their inventive "Self."

Denmark is known for its excellence in design. Their design is very much sophisticated. But on the other hand, they show us that there are other designs. For example, Friz Hansen is famous for their chairs. They produce excellent quality artificial leathers for their chairs. But they sell chairs with natural leathers with scratches at a higher price. Why?

Their answer was, "You can enjoy stories. You can enjoy talking about how the animal may have lived and how the scratches were made." It was a big surprise. I thought they sell products, but they are selling processes. They know how process values are important.

So, sophisticated design represents Danish culture on one side, but Lego represents another culture of Denmark. They really know what design is. Further, Denmark is known for Vikings. Vikings are explorers. We can understand why Danish design is exploring many frontiers.

### *3.5.2 Adaptive Model*

Another way to revive production with limited number of facilities and equipment after World War II was model upgrading. As product variation cannot be developed, the same model is produced but it keeps upgrading.

This satisfied customers very much. Deci and Ryan pointed out in addition to their self-determination, the importance of growth for humans. Customers enjoyed imagining the next upgraded model. And indeed, it satisfied their need for growth.

### *3.5.3 Continuous Prototyping*

Deci and Ryan also pointed out our need to grow in addition to their famous Self-Determination Theory. In fact, growth is another adaptation. We grow by adapting ourselves to the environments. Evolution is another adaptation, and Evolution and Growth are deeply associated.

The idea to satisfy our needs for growth was introduced in software in the early twenty-first century. This approach is called, "Continuous Prototyping" (Fig. 3.11). In this approach, very basic functions are first provided. After customers get used to and become confident, then software developers provide little higher functions. Thus, as the customers get used to and become confident, the functions are upgraded, and higher functions are provided step by step.

We should take note that confidence and trust are basically the same. Confidence means trust in yourself and Trust means trust in others. In fact, in German, Confidence and Trust are expressed by the same word "Vertrauen."

Therefore, the customers' trust in the system increases with the increase of their confidence. And software developers have their own benefits, too. They can sell a system at an early stage of development, and they can understand what their customers expect by observing their use. In other words, they noticed the importance of processes. By introducing such an approach, software developers succeeded in satisfying our self-actualization needs.

But hardware development keeps the old tradition of delivering the finished product to customers.

So, modularization may be understood to cope with the diversifying customer requirements with limited facilities and equipment (Fig. 3.12).

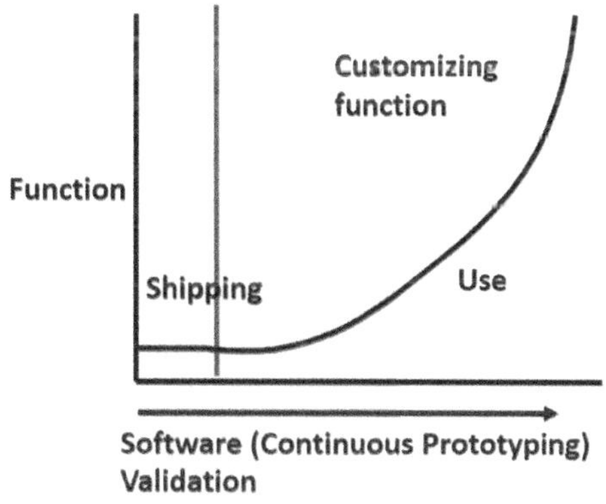


**Fig. 3.11** Software development

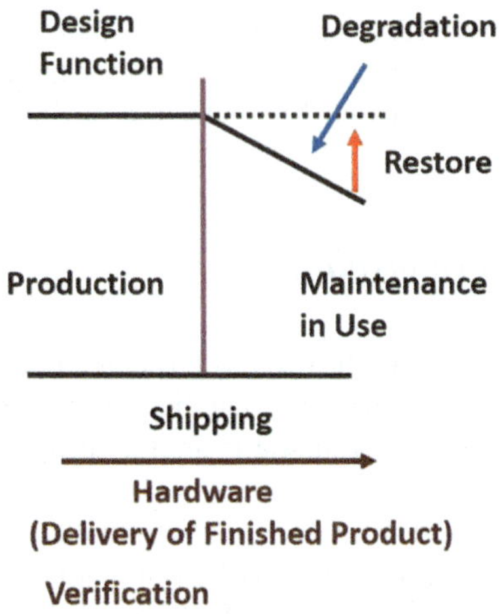


**Fig. 3.12** Hardware development

### *3.5.4 Degradation: Look Upside Down*

The invention of a new programming language which can be used in common at the stage of development and for applications led software development industry to the introduction of "Continuous Prototyping" development style. This success opened the eyes of hardware industry how important processes are. However, they cannot but stick to their traditional fixed function development (Fig. 3.12).

But if we turn the software development (Fig. 3.11) upside down, then it is nothing else than the degradation of hardware (Fig. 3.13). And this reminds them of the advantage of repair or the advantage of utilizing deterioration. In fact, degradation is the opposite of growth.

### *3.5.5 Fitting, Breaking In*

Asics, shoes maker, realized that degradation is associated with the feeling of fitting. When the shoes deform as shown in the middle, people throw them away. But Asics thought this is the form of the shoes that fits walking best. So, they developed shoes with flexible middle part. Then, people who wore them said, "I feel like walking on my own feet. I can enjoy walking, no matter how much long it may be." These shoes sold very well. And Asics applied the same idea to the sport shoes, because athlete shoes deform in the same way at the start. This idea was highly evaluated by athletes, too. This is another example of adaptability.

To increase adaptability, we need to look at things from a different perspective. And again, this is another expression of "Self" (Fig. 3.14).

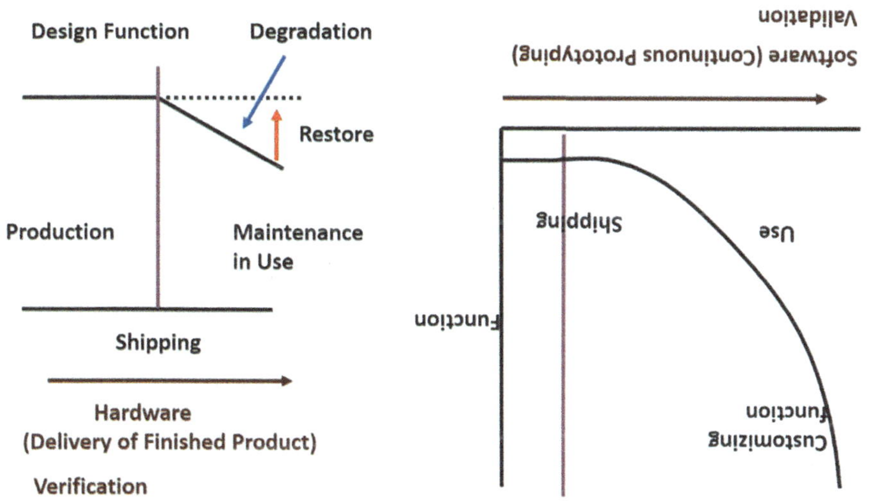


**Fig. 3.13** Degradation is reversed growth

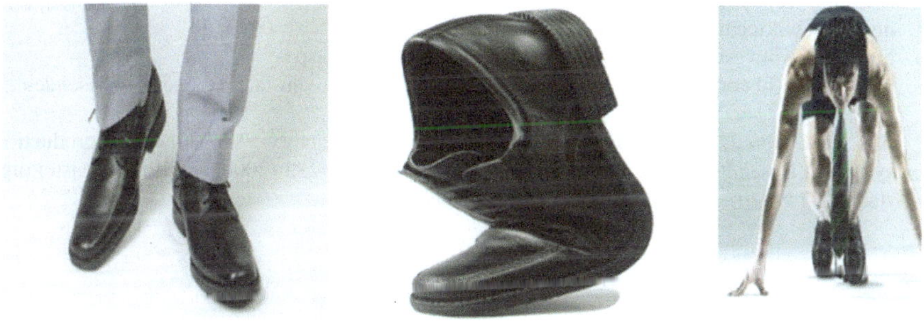

**Fig. 3.14** Asics shoes

## 3.6 Increasing Importance of "Self"

We have seen how "Self" is increasing its importance. But we should be careful. We are not discussing Personalization. Indeed, our characters and behaviors vary widely from person to person. But this is the problem of personalization. When industry talks about diversification and personalization. They are focusing on such individual diversity.

The diversity of "Self" comes from our natural need to expand the community of human species wider and to survive longer. That is why "Self" becomes most important in Maslow's hierarchy of human needs. It is the need of the human species. Not an individual want. That is why "Self" determination is important in Deci and Ryan's SDT. These are our efforts as the human to expand our community and to survive longer.

We also must be careful that the word "self" was used in autonomous distributed system in such a way as "self-regulating," etc. But in these cases, machines operate as they were expected, and no special instructions were needed from us. But how they operate was strictly regulated by us. It is nothing other than just skipping the step of giving instructions.

The "Self" here is our efforts as the human species. We are born to do such efforts as a living thing. Engineering looks completely artificial, but it is not. It is one of our efforts to achieve this goal. That is why we are called "Homo Faber." We make tools to control our environments to satisfy this need as human species.

## References

1. A.H. Maslow, A theory of human motivation. Psychol. Rev. **50**(4), 370–396 (1943)
2. https://en.wikipedia.org/wiki/Tertiary_sector_of_the_economy
3. http://art.yale.edu/file_columns/0000/1474/homo_ludens_johan_huizinga_routledge_1949_.pdf
4. E.L. Deci, R.M. Ryan, *Intrinsic Motivation and Self-Determination in Human Behavior* (Plenum, New York, 1985)
5. https://en.wikipedia.org/wiki/Self-determination_theory
6. https://www.investopedia.com/terms/m/masscustomization.asp
7. https://www.history.com/this-day-in-history/volkswagens-dream-factory-opens-in-resende-brazil
8. A.R. de P. Abreu, H. Beynon, J.R. Ramalho, 'The dream factory': VW's modular production system in Resende, Brazil. Work Employ. Soc. **14**(2), 265–282 (2000). https://www.jstor.org/stable/23747870?seq=1#page_scan_tab_contents

# Chapter 4
# Mind-Body-Brain

Changes yesterday were smooth, so we could predict the future. Therefore, engineers could foresee the operating conditions, so their efforts were paid to develop final products with better functions. Functions and their reproducibility were most evaluated in those days.

But changes are sharp these days, so we cannot predict the future. Further, our world changed from a closed world with boundaries to an open world without boundaries. And this expansion is accelerating rapidly. Thus, the degrees of freedom increase tremendously. So, we need to make decisions immediately to adapt to the unpredictable changes of the environments and situations.

Design has been discussed with chief attention paid to functions. Engineering design so far has been carried out this way. But design is creation. We make decisions on what new world we will create. So, we have to remember design is decision making.

Engineering design up to now had a clear goal. So, engineering design was problem solving. But as environments and situations come to change in an unpredictable manner, "Adaptability" becomes most important. But to adapt to such unpredictable changes, we need to be highly situational aware and perceive and understand the environment and the situation exactly.

Design yesterday may be interpreted as a human species endeavor to expand human community in the future, because humans seem to be the only living animal that creates the future. But today, humans have to adapt to the current changes as all living things do. "Adaptability" is not human-specific, but all living things need to survive.

Therefore, design is getting closer and closer to decision making. In this sense, all living things are designing their way to adapt and to survive. This is nothing other than evolution. Adaptability and evolution are the same.

When we ask somebody to decide, we use the word, "Make up your mind." Well, then what is mind?

 
S. Fukuda, *Self Engineering*, SpringerBriefs in Applied Sciences and Technology,
https://doi.org/10.1007/978-3-030-26725-4_4

Mind-Body-Brain Problem is often discussed. When we say "Mind," most people associate it with the brain. The brain plays a very important role to make judgments and decisions in the case of vertebrates. We believe that our brain is full of knowledge so that it will give a solution how to cope with the environment and situation.

But we should remember that knowledge is the experience of the past. Many pieces of past experiences are passed on from generation to generation and are stored and structured them into knowledge. When the environments and situations change in a predictable manner, knowledge indeed worked very well. But today, changes are unpredictable, so we cannot rely on past experience or knowledge anymore.

Let us think about the octopus. The octopus is the typical invertebrate and is positioned to the opposite side of the human in evolutionary tree. The octopus parents die very soon after their babies are born. So, experience of the previous generation is not passed on to the next generation. Thus, the octopus cannot help but live only on their instinct.

But the octopus is known as Master of Escape. They can overcome almost any difficulties. A typical example is they can escape from a screwed container. They find how they can escape by trial and error. They do not have any prior knowledge. How about us, humans? We may be panicked, and cannot find any solution, because we do not have such experience before.

If we compare the octopus and the human (Fig. 4.1), we immediately realize why the octopus is classified as Cephalopod (Greek head + foot). In English, their foot is called arm, and indeed their foot (arm) is very soft and multipurpose. Thus, for the octopus, "Mind" is body. What the octopus teaches us is the importance of direct interaction with the outside world.

As pointed out in the above, "Mind" is associated with decision making. In the case of the octopus, they look for the solution by trial and error by directly interacting with the outside world. Their mind is body, and it works on the peripheral. The octopuses have brains, too. But since they do not transfer knowledge from generation

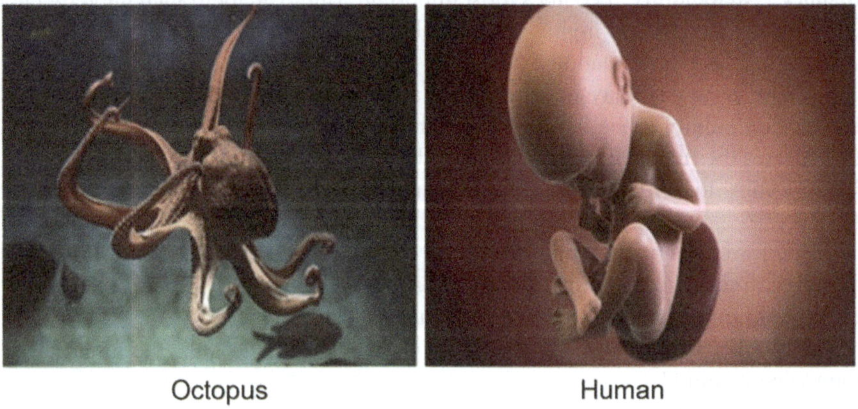


**Fig. 4.1** Octopus and human

to generation, their brains do not provide "Knowledge." It is better to say they work to produce "Wisdom."

Peter Godfrey-Smith, Professor at the University of Sydney, published "The Other mind: The Octopus, The Sea and Deep Origins of Consciousness" [1, 2]. He works primarily in philosophy of biology and mind. And he dives very often and observes the octopus. He is also known for his Pragmatism. This book describes in great detail how differently intelligent the octopus is. We feel there are many things we learn from the octopus.

John Z. Young, who is called "one of the most influential biologists in the twentieth century" and studied brain [3], points out that we should take note of another intelligence. It is not transmitted as knowledge from generation to generation, but is transmitted as genes. As unpredictable changes take place frequently and extensively, we should pay attention to such intelligence transmitted in the form of genes, because it does not take time to adapt and it enables very flexible response.

We should regard evolutionary tree from another perspective. It describes different ways of developing intelligence. Vertebrates, the typical example of which is the human, must learn many things from invertebrates, the typical example of which is the octopus. Jacques Cousteau, French researcher who studied the sea and all forms of life in water [4], called the Octopus "The pirate of the Sea."

In the case of the human, we have many minds inside of us in addition to such system as peripheral nervous system. Brain constitutes central nervous system, but before reaching the brain, many nervous systems are working. They connect the limbs and the organs to the brain. Some external stimuli are processed before they are sent to the brain. The reflex is one of such typical examples. We must realize we need to coordinate and balance all of our body parts. Researches are progressing rapidly in each system, but how we coordinate and balance all of these sensors and actuators of our body is still unknown (Figs. 4.2 and 4.3).

Human movement control is divided into two. One is called "Motion Control," which is to control our movement from outside. The other is called "Motor Control." Although we succeeded portion of Motion Control as we see today, but as Nikolai Bernstein points out human motion trajectories vary widely from time to time. But when we come close to the target, trajectories are fixed (Fig. 4.4). Therefore, he points out coordination is important in human motion control [5, 6].

**Fig. 4.2** Mind-body-brain

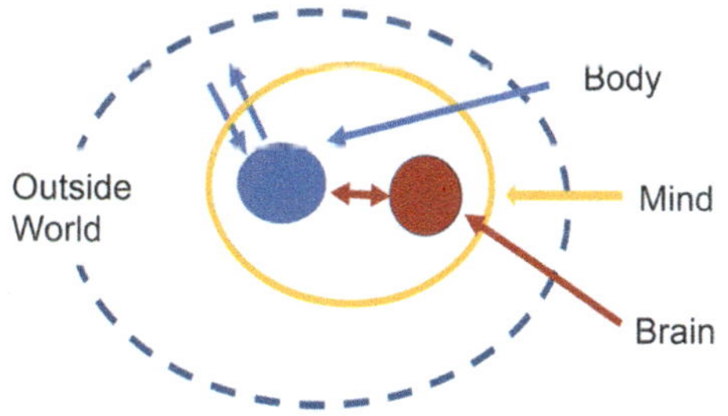

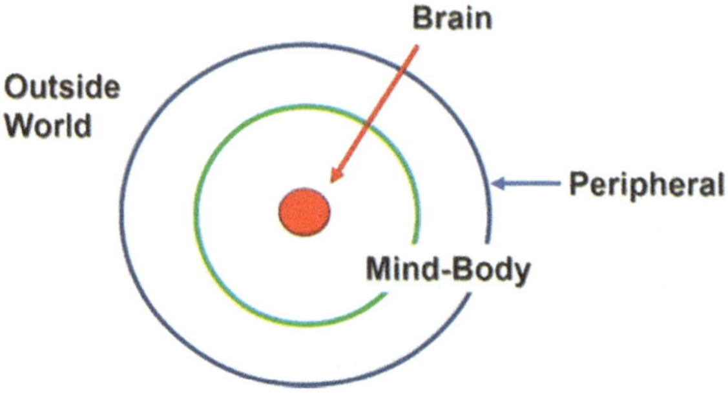


**Fig. 4.3** Mind-body-brain

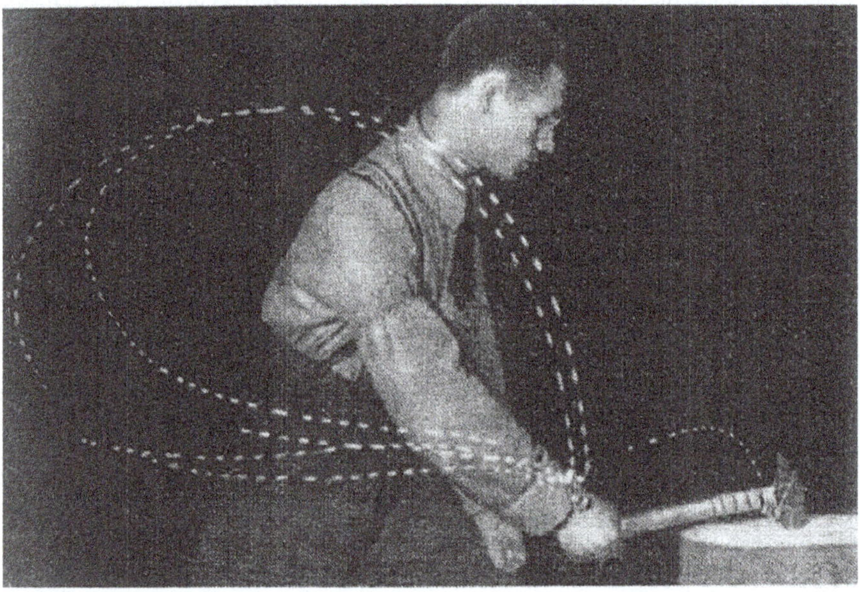

**Fig. 4.4** Cyclogram of hammering

Although he did not point out, but recent physiological studies are making clear that our muscles harden when we realize we are getting close to the target, so that our trajectories are fixed and reduce the degrees of freedom to the minimum. Most human motion control is focused on this stage alone, and machine motion control is developed based on this last stage trajectory, because it is reproducible and can be expressed in a rational and explicit way.

Although part of Motion Control is being carried out successfully, its whole is still not made clear. Well then, what about Motor Control. It is no exaggeration to say all is still unclear, although particular researches are progressing very well.

Humans have five senses as is well known. But we have additional senses such as nociception (pain); equilibrioception (balance); proprioception and kinesthesia (joint motion and acceleration); sense of time; thermoception (temperature differences); and possibly an additional weak magnetoception (direction). In addition to these numerous sensors, we have numerous actuators, so that we do not know how we are coordinating and balancing them.

Indeed, this is the tacit and nonverbal world.

As our movement trajectories indicate, it may be safely said that we are coordinating and balancing our body parts by trial and error. And it should be added that this is true "Self"-control. It is not the traditional self-control, which is just responding to the instructions already given as we have seen in machine control.

And we must remember Herbert Simon pointed out rationality is bounded. If the degrees of freedom increase very much, we cannot optimize rationally and we can't help but satisfying ourselves with the solution which satisfied us enough. Thus, he proposed Bounded Rationality [7] and Satisficing (Satisfy + Suffice) [8, 9].

Indeed, think of global optimization using simulated annealing [10] (Fig. 4.5). We iterate many times, and we believe the highest peak after so many repetitions is the optimum. But there is a possibility that the next iteration might provide the higher peak. So, we just satisfy ourselves by believing that the highest peak so far is the highest because we iterated so many times. So, if the search space becomes tremendously large, we cannot truly optimize.

And in addition to the problem of tremendously large search space, we have limited time to make decisions. Gerd Gigerenzer and Heinhard Selten edited the book "Bounded Rationality: The Adaptive Toolbox" [11]. Such cases of decision making how we judge the patient condition and which emergency hospital or which medical department we should carry him to. We do not have time and our information is limited. We must decide immediately.

Simon and Gigerenzer just pointed out that rationality is bounded. Simon proposed Satisficing, but he did not tell us how we can satisfy ourselves. This book is written to provide a tool how to satisfice us.

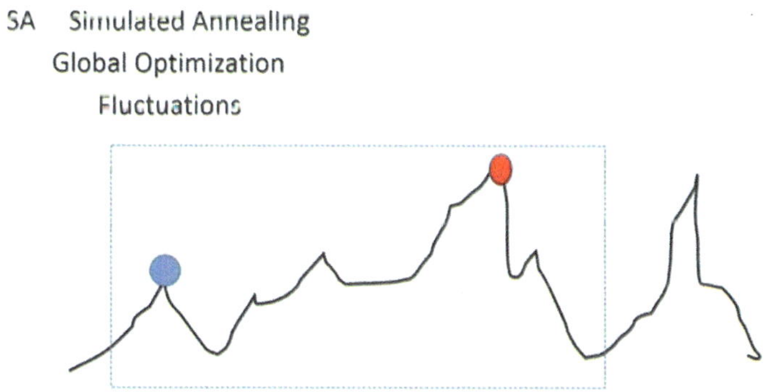


**Fig. 4.5** Global optimization

As the world is expanding very rapidly and it is shifting very quickly from explicit and verbal to tacit and nonverbal. Our world now is nothing else but the world of tacit dimension as Michael Polanyi pointed out [12].

## References


1. https://en.wikipedia.org/wiki/Peter_Godfrey-Smith#cite_note-8
2. P. Godfrey-Smith, *Other Minds: The Octopus, The Sea and Deep Origins of Consciousness* (New York Farrar, Straus and Giroux, 2016)
3. https://en.wikipedia.org/wiki/John_Zachary_Young
4. https://en.wikipedia.org/wiki/Jacques_Cousteau
5. https://en.wikipedia.org/wiki/Nikolai_Bernstein
6. N.A. Bernstein, *The Co-ordination and Regulation of Movements* (Pergamon Press, Oxford, 1967)
7. https://en.wikipedia.org/wiki/Bounded_rationality
8. https://en.wikipedia.org/wiki/Satisficing
9. H.A. Simon, *Administrative Behavior: A Study of Decision-Making Processes in Administrative Organization* (Macmillan, New York, 1947)
10. https://en.wikipedia.org/wiki/Simulated_annealing
11. G. Gigerenzer, R. Selten, *Bounded Rationality: The Adaptive Toolbox.* (MIT Press, Boston, 2002)
12. M. Polanyi, *The Tacit Dimension* (University of Chicago Press, Chicago, 1966)

# Chapter 5
# Emotion

Emotion is defined in many ways. Fukuda defines "Emotion" [1–3] as follows.

"Perception (Cognition) → Motivation → Decision Making → Action → Emotion" cycle as shown in Fig. 5.1.

This definition derives from the fact that emotion and motivation came from the same Latin word "movere (=move)." And e=ex(=out). So, "Emotion" means to move out. You take action to the outside world, based on your understanding of the environment and situation, and your motivation.

This definition is unique. Most definitions consider "Emotion" as a result. But Fukuda's definition puts its primary focus on perception and decision making. In other words, it places "Emotion" on the starting point.

Such definition is very few. Why Fukuda defined "Emotion" in such a way is because the way changes take place has changed. Yesterday, changes were smooth, so that we could differentiate them and predict the future. But today, changes are sharp, so we cannot differentiate them. In other words, we cannot predict the future (Fig. 5.2).

Thus, yesterday, engineers could foresee the operating conditions of machines. So, they could focus on functions of their final products. Therefore, functions, their reproducibility and robustness were most important. But today, they cannot. So, "Adaptability" becomes most important. We should remember "Evolution" is "Adaptation." It is adaption as "Species."

Maslow proposed "Hierarch of Human Needs" in 1943 [4], Fig. 5.3. Why Maslow's hierarchy is getting wide attention these days is because it shows that at its low level, we make efforts on individual needs to survive. "Self-Actualization" is at the highest level because it is what we do for human species to adapt and to survive.

AI is drawing attention again these days. Deep-Learning is the basic tool for today's AI. Deep-Learning-based AI works excellent for games with perfect information. And the game with perfect information can be solved with backward induction. In short, we can start from the goal and reasoning backward.

 
S. Fukuda, *Self Engineering*, SpringerBriefs in Applied Sciences and Technology,
https://doi.org/10.1007/978-3-030-26725-4_5

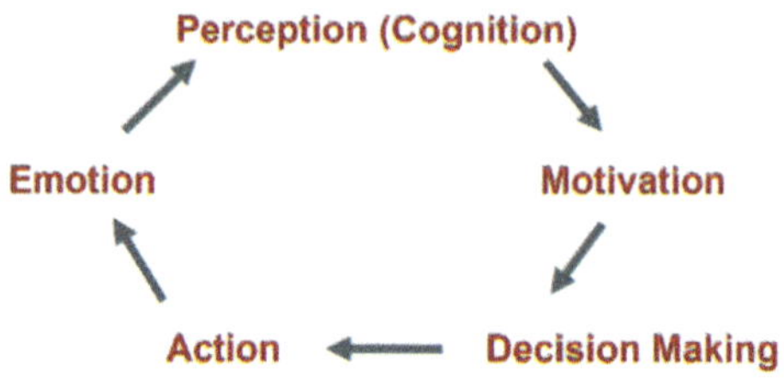


**Fig. 5.1** Emotion cycle

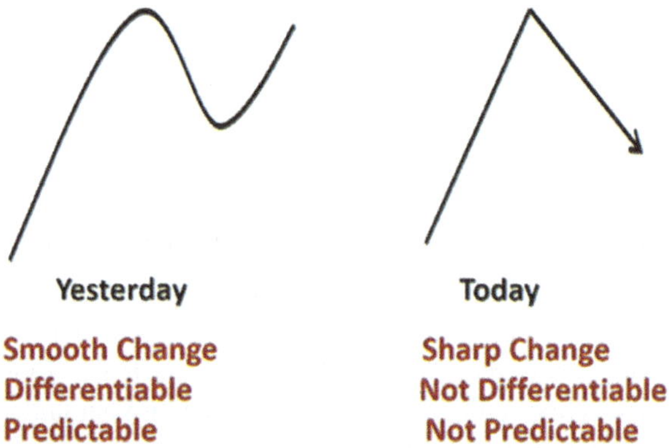


**Fig. 5.2** Changes of yesterday and today

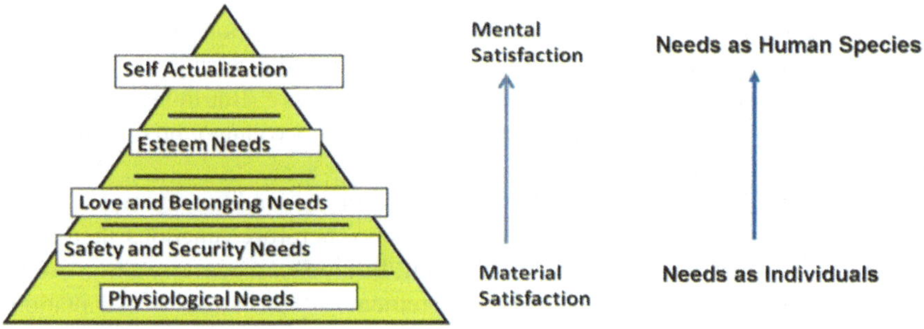


**Fig. 5.3** Maslow's hierarchy of human needs

In such a case as sports, the number of players is fixed and there are strict rules. So, it is a typical game of perfect information. But in other fields, too, the number of people involved is almost fixed and there are rules, so Deep-Learning-based AI is expected to contribute much to the current industrial or organizational framework. It is the explicit and rational world. So, members can converse using symbols.

But we have to remember that the real world is more complex and complicated. Traditional Engineering is making efforts more and more to establish the well-defined world. We pay tremendous efforts to change ill-defined things to well-defined. But no matter how much we may pay efforts, the real world is ill-defined. We must accept this fact. The most important role for humans is to make decisions. Humans are most valuable as decision-makers. Even if we make up a team with machines, and humans and machines work together as team members on the same team, what is important is the final decision. We should not forget that it is us, humans, who make the final decision.

But humans coordinate and balance our body parts and move smoothly. Engineers try to automate processes and in order to do that, they make tremendous efforts to represent them in an explicit and rational way. But if we, engineers, leave the decision to humans and we prepare some supporting tools for their decision making, that would be far easier and does not cost or time. And for humans to move in the real world, they do not need all information. Well, sometime, they may need, but such a case is very few and the problem itself is small.

Coordination and Balancing involves strategy and tactics. Humans find strategy and tactics by trial and error, because every time they move, they are creating that movement. For humans to move, it is the act of creation.

As discussed above, Machine Learning and Data Mining are getting wide attention with the progress of AI. But for humans to coordinate and balance their bodies and *their thoughts*, what they need is not data, but information. They need multisensory information to understand the real world to make adequate decisions. And often they do not need all information. They need information for their decision making.

And with the progress of IoT, how humans and machines learn together becomes important. It is a challenge to establish the approach how they can learn together. Such team working is the problem of organizing an adaptive network.

Why "Graph Theory" is increasing importance is one reason is this is the mathematical ground for organizing adaptive networks. The other reason is we can understand the whole at once. We do not need to follow the mathematical reasoning, but we understand the situation at a glance.

In other words, network representation is nothing other than pattern representation. We can understand the situation at once, if it is represented as a pattern. That is why the pattern approach is introduced in Mahalanobis Taguchi Fukuda approach.

In AI and in other engineering fields, the importance of "Pattern Recognition" is emphasized. But we should be careful. The pattern approach introduced by Fukuda is "Pattern Cognition." "Pattern Recognition" is the Pattern Classification of data based on knowledge. That is why "Pattern Recognition" is emphasized in Machine Learning. Its process is automated. Knowledge Discovery Database (KDD) is the same.

The word "Pattern" used to mean "Patron" in Middle English, which means "Model." So, patterns do not have to be related with knowledge or data. Fukuda emphasizes "Pattern Cognition," although this may be the word he coined.

Cognition is understanding through thought, experience and the senses. It is understanding through perception, intuition and reasoning. We need to learn more from the octopus. They are representing the invertebrate in Evolution Tree. But they solve

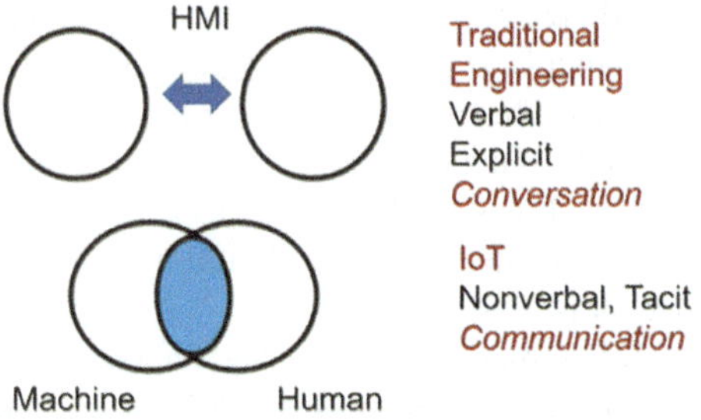


**Fig. 5.4** From conversation to communication

the problem very intelligently. They use their "Instinct." It is their "Wisdom." Not "Knowledge."

Pattern Cognition needs to be processed by trial and error. Therefore, learning from failures works very well for pattern cognition. When you get excited and motivated, you feel your heart beating, your body getting hot, etc. This is our preparation. We perceive and get motivated and prepare ourselves for action.

Pattern in "Pattern Recognition" is Signal or Symbol. Pattern in "Pattern Cognition" is Sign. So, "Pattern Recognition" and "Pattern Cognition" may be compared to "Conversation" and "Communication."

As the products come to be designed and produced as a team, communication becomes critically important. In conversation, we can exchange symbols (words) and we can process the job. But team working needs more than words. We need to share our feelings. The listener needs to understand what is on the speaker's mind and what he would like to tell before he puts it into words. Figure 5.4 shows the case of Human–Machine Interface, but the same holds true with Human–Human interaction.

Patterns in "Pattern Cognition" may be described as our attempt to represent our search for appropriate model or for appropriate mind and its processes as patterns.

This book emphasizes the importance of humans as decision-makers. We should coordinate and balance information to make adequate decisions as we do to control our movement. We should realize what capabilities we are born with. In other words, the role of "Instinct" is increasing.

## References

1. S. Fukuda, *Emotional Engineering*, vol. 1–7. (Springer, London, 2011–2019)
2. S. Fukuda (ed.), in *Advances in Affective and Pleasurable Design*. Proceedings of the AHFE 2018 International Conference on Affective and Pleasurable Design, Advances in Intelligent Systems and Computing, vol. 774. (Springer, 2018)
3. S. Fukuda (ed.), in *Advances in Affective and Pleasurable Design*. Proceedings of the AHFE 2019 International Conference on Affective and Pleasurable Design, Advances in Intelligent Systems and Computing, vol. 952. (Springer, 2019)
4. Abraham H. Maslow, A Theory of Human Motivation. Psychol. Rev. **50**(4), 370–396 (1943)

# Chapter 6
# Experience Yesterday, Today, and Tomorrow

## 6.1 Yesterday

### *6.1.1 Experience Was a Thing of the Past*

There were no appreciable changes yesterday in environments and situations. Tomorrow was another today. So, experience played an important role. Experience was stored and structured into knowledge. And we paid a lot of efforts to pass knowledge from generation to generation. That is education. Education was important for deepening and widening knowledge.

Engineering was no exception. As industry was individual-product-centric, experience played an important role in designing and producing better products. And our world was small and closed with boundaries. So, our daily life sphere was very much limited. Therefore, "think in the same framework" and "share the same perspective" were important. There was no room for "Self."

The fact that the first buzzword in Artificial Intelligence (AI) was "expert systems" demonstrates how experience was important. How we can make the best use of the experience of experts was the focus of AI. And we should also remember that the etymology of experience and expert is the same Latin word "experiri (=try)."

Mycin, which was developed at Stanford in the early 1970's, was the first, well-known expert system. It aimed at identifying bacteria causing severe infections [1, 2]. And the word "Knowledge Engineering" [3] became very popular together with the word "Expert Systems." Although the word "Expert Systems" disappeared soon, because many pieces of expert knowledge are specific to that field and cannot be applied to other fields and the real world is far more complicated and complex. "Knowledge Engineering" survived, because it is a tool applicable to all fields.

The etymology of experience is the Latin word "experiri" (=try). And experiment came from the same Latin word. So, experience was an action to explore your own world and make your knowledge richer. But experience was and has been regarded as a thing of the past.



S. Fukuda, *Self Engineering*, SpringerBriefs in Applied Sciences and Technology,
https://doi.org/10.1007/978-3-030-26725-4_6

No. It is not. That is what I emphasize in this book. Experience is our experiment toward the future. Such misunderstanding about experience put restraints on our action.

## 6.2 Today

### *6.2.1 Concept Dominating World*

Betty Edwards [4, 5] published a best seller "Drawing on the Right Side of the Brain," in which she made clear that we draw the real world until seven years old. But after this age, we come to look at things as concepts.

Child under seven years old draws a car from the front. So, it is difficult for us to know whether he or she draws a human face or a car. But after seven, he or she draws a car from the side. This may be associated that our eyes move sideways easier, but anyway, we can understand easily that he or she drew a car.

Antoine de Saint-Exupery published "the Little Prince" in 1943 [6]. In this book, the little prince drew the scene of a large snake swallowing an elephant. But every grownup who saw the drawing said, "It's a hat." The little prince drew the real world.

We will know easily how we are obsessed with concepts, if we do the following simple experiment. Give a talk in a room. And ask everybody to get out of the room. Then, ask them if there was a clock in the room. About half of them do not remember whether there was or not. The other half remember there was a clock. Then, ask this other half about its details, shape, dial plate, etc. Then, only one or two can answer exactly. Others recognized the clock, but they remember it as a concept. There was something that tells time. They look at the clock only in terms of its functions.

When we say "Experience," more often than not, we only think from the viewpoint of concepts. Yesterday, when the environments and situations did not change appreciably, the real world and concepts were identical. So, concepts worked effectively. This is why we could and have been able to communicate using symbols. Or to be more exact, we conversed using symbols.

### *6.2.2 Understand the Real World Exactly to Adapt*

But today, environments and situations change frequently and extensively, and these changes are unpredictable. We cannot know what will happen next. Thus, "Adaptability" emerged as most important. To adapt, we need to be situational aware and perceive the environment and the situation correctly. We cannot make decisions and act on concepts anymore. We need to understand the real world exactly. Then, we can make appropriate decisions and take actions.

"Emotion" (Chap. 5) is often taken as our response to the outside stimuli. But this interpretation is not adequate. We all know how our hearts and bodies get excited to adapt to the changing situations. Emotion is, thus, a trigger or preparation to the action to adapt ourselves to the changing situations appropriately. "Emotion" and "Motivation" come from the same Latin word "movere=movere."

Emotion is e=ex=out+movere=move. So "Emotion" means to move out. You are motivated and you would like to take action to the outside world to satisfy your motivation. At least, this is the role of emotion and the need to understand the real world correctly is increasing, so emotion is increasing its importance. And this is the phenomenon observed in vertebrates.

As discussed in Chap. 4 "Mind-Body-Brain Problem," invertebrates interact with the real world in different ways. These may be called Natural Wisdom. It is not knowledge at all. But we must remember we are also born with such capabilities or with such an instinct to adapt to changing environments and situations, although in a very much different way from invertebrates. It is our wisdom.

### 6.2.3 Sharing Experience for Team Strategy

Yesterday, engineering is individual-product-based. But today, products are designed and produced as teams. Thus, experience has been discussed on an individual basis until today, but today not only machines, but we also work together as a team to cope with the rapidly increasing complexity and complicatedness. Therefore, we must consider how we can share experience among team members. Sharing experience among human team members is an imminent task today.

Knute Rockne, famous American football player and coach, left a famous word "11 Best, Best 11." He pointed out that the best team cannot be made by 11 best players. To make up the best team, 11 players who understand the strategy and work together for the team are needed." He demonstrated by bringing University of Notre Dame to the ever-winning team.

As I used to be a soccer player, let me explain by taking soccer as an example. Yesterday, players were expected to do their best in their own positions, because situations did not change during the game, so the formation did not change. But today, situations change frequently during the game, so now midfielders become playing managers.

Yesterday, managers could give instructions from outside of the pitch. But today, only players know the current situation. As a midfielder can understand the flow of the whole game, he became a playing manager. Now a manager is on the pitch (Fig. 6.1).

When the manager was outside of the pitch, he can give instructions in words to players. Now, he is on the pitch and he is playing himself. So, there is no time to give instructions and if he does, the enemy team understands the next action. So, the players need to understand what the playing manager is expecting from them and to prepare for the next strategic expansion.

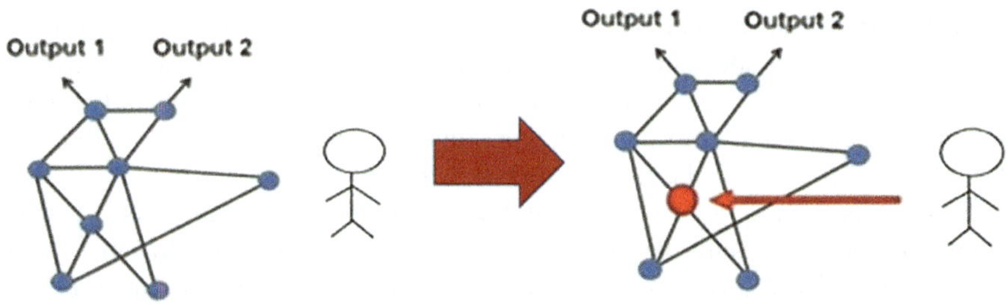


**Fig. 6.1** Outside of the pitch to on the pitch

In other words, yesterday manager and players can converse, but today they need to communicate. Communication is reading mind. Be careful not to confuse with mind reading AI, which is attracting attention these days. It would be better to say, reading intent. The listener estimates what the speaker has at the bottom of his heart.

But although it is nothing else but strategy, we should be careful. This is "Bounded Strategy." In sports such as soccer, the number of players is fixed and there are strict rules. And in other organizations, too, the number of team members is roughly fixed and there are rules. So, game or organizational strategies are choice of strategies within these constraints. Indeed, it is how we adapt the network to the environments and situations, but this network does not change. Only edges are linked in different ways. So, this is the game of perfect information. Therefore, the current AI and Deep Learning will be very effective. It will produce strategies, which we fail to notice.

### *6.2.4 Sharing Economy: Broaden Your Perspective*

"Sharing Economy" is getting wide attention these days. The same car is used in different ways by different drivers. Passenger cars could enjoy long sleep in the garage yesterday. Indeed, cars played a role of status symbol more than for driving. But if we broaden our perspective, we find out we can enjoy cars more, if we let others use our own cars. Probably cars themselves are enjoying their life, going through many different experiences.

"Sharing Economy" may be interpreted as our wisdom to make the most of our resources and reduce waste. But we also must remember that it is achieved by broadening our perspective. Perspective broadening is nothing other than creation of the new world. It satisfies our human need to self-actualization. So, in fact it is deeply related to "Self."

## 6.3 Tomorrow

Tomorrow, we need to face the real world and tackle with it. As environments and situations change unpredictably, what we need is to come back to the original definition of experience (=experiment), as the Latin etymology experiri (=try) indicates. We need to "try" to find out whether the strategy on our mind right now works or not.

Experience has been a thing of the past for a long time. But from today, we need to change our perspective. We have to create an experience to find out if the strategy works or not. Thus, experience tomorrow is a thing of the future. And the current AI and Deep Learning will not work anymore. We need to develop another AI framework.

Therefore, what we can do is to find a strategy by trial and error. The current AI is very effective when the goal is clearly defined. But tomorrow we need to find a goal and how to get there by trial and error. This is the world of Pragmatism. We must learn by failures. We need to develop another strategy without any constraints, if the outcome fails our expectations. In other words, we need to develop a network truly adaptive to the real world from scratch.

## References

1. https://en.wikipedia.org/wiki/Mycin
2. B.B. Buchanan, E.H. Shortliff (eds.), Rule-based expert systems, in *The MYCIN Experiments of the Stanford Heuristic Programming Project*. (Addison-Wesley, Boston, 1984)
3. https://en.wikipedia.org/wiki/Knowledge_engineering
4. https://en.wikipedia.org/wiki/Betty_Edwards
5. https://www.drawright.com/
6. https://en.wikipedia.org/wiki/The_Little_Prince

# Chapter 7
# Experiments

## 7.1 Rational Experiments

As described in the previous chapters, the etymology of experience and experiment is the same. Both came from the Latin word "experiri (=try)." Just like experience has been regarded as a thing of the past, experiments have been often regarded as a tool to "visualize" the fact. But this is science. In science, what matters is truth. Experiments in science are to find a truth or to confirm it.

But experiments in engineering are for decision making. It is to find out whether the hypothesis works or not. So, the goal is not fixed. It depends on whether the route is easy or not. More often than not, we need to change the goal, so that we can get there more easily. In fact, such searches themselves satisfy our "Self."

Engineering is to make our dreams come true. But how we make our dreams come true is up to us and if it takes too much of efforts, then we will change the goal and will look for another route which provides better satisfaction. So, the goal and route in engineering are very flexible, and they can vary with trials and errors.

In other words, science is rational, but engineering is very much pragmatic. Traditional engineering has pursued rational and explicit approaches, too. Their idea was the same as identifying the name of a river (Fig. 7.1). If we look at the river itself, we cannot identify its name. River is flowing and we cannot predict its behavior. But if we look around and find mountains, forests, etc., that does not change, then we can identify its name.

System identification is the same idea. We enter the known input and observe the output and identify the system, and we know what parameters we should process (Fig. 7.2). This idea contributed to expand the rational world. When the world was small and closed with boundaries, we could easily find rational and explicit solutions. But as the world expanded and became an open world without boundaries, such straightforward way did not work anymore. But by introducing system identification, we could expand the rational world and establish a controllable world (Fig. 7.3).



S. Fukuda, *Self Engineering*, SpringerBriefs in Applied Sciences and Technology,
https://doi.org/10.1007/978-3-030-26725-4_7

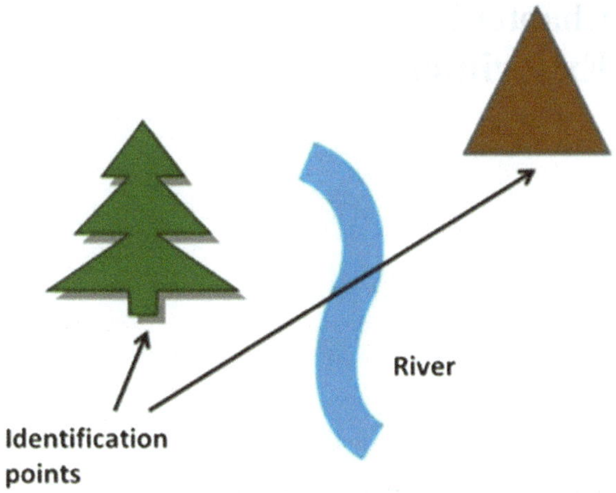


**Fig. 7.1** Identifying the name of a river

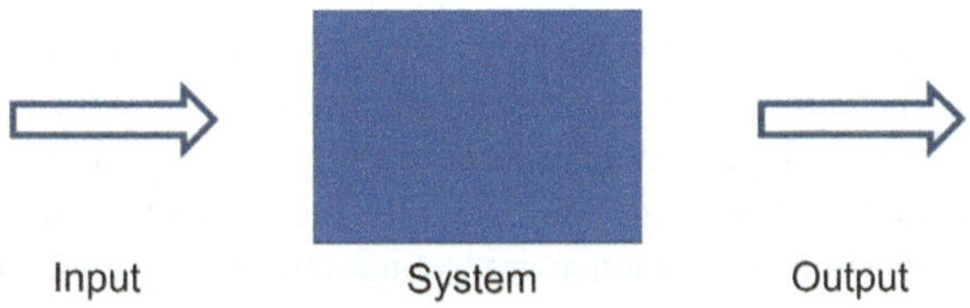


**Fig. 7.2** System identification

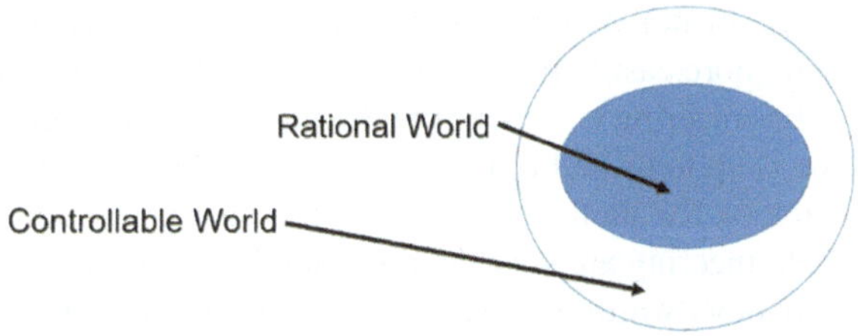


**Fig. 7.3** Rational world expanded to controllable world

## 7.2 Pragmatic Experiments

But the expansion of our world was so rapid that such rational approaches soon came to be invalid and we need to change our strategy to pragmatic experiments, i.e., trial-and-error approach.

Many pragmatic approaches have been proposed. But here, let us take Walter A. Shewhart's PDSA as an example first [1] (Fig. 7.4). First, you think of a hypothesis (PLAN). Then apply it to the problem (DO). Then, study if it works (STUDY). If it does, that's fine. If not, apply another hypothesis (ACT). Repeat this cycle until you find the working hypothesis.

**Fig. 7.4** PDSA cycle

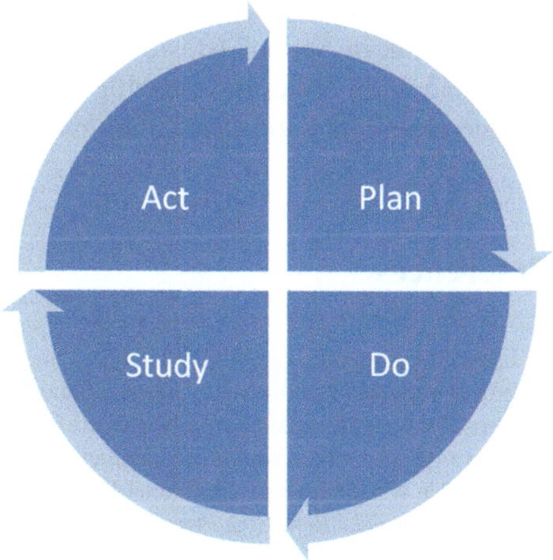


PDSA is known as Design of Experiments. But the meaning of experiment is very much different from the rational experiments. First of all, this is totally processed by trial and error. And even when your hypothesis does work, it does not mean there are no other hypotheses that work better than the current one. What you can say is you found the path to the goal. That's all. There might be far better paths to the goal. What matters here is as Shakespeare said, "All is well that ends well." It does not say "ends best."

As environments and situations change frequently and extensively in an unpredictable manner, it is most important to find the path to the goal. We need immediate action to adapt to them. Pragmatism is the soul of exploration. What is important is we can find the path to the goal, no matter how much good it is. If we can find a path, then we can move forward, no matter how much small it may be.

It is interesting to know that Shewhart self-learned statistics, although he has rich knowledge about physics and mathematics. He must have realized the limit of rational and explicit approaches. He proposed this idea in the mid-twentieth century, when mass production was prevalent. He was seeing the future.

Far before Shewhart, Charles S.Peirce proposed "Abduction" [2–4]. He is called "the Father of Pragmatism." His abduction theory runs just the same way as PDSA. You "abduct" an idea from somewhere and apply it to the current problem. Study if it works. If it does, then it is fine. But if not, repeat the process until the working idea is found. He proposed "Abduction," when the westward movement was at its peak in mid-nineteenth century.

As the USA is a young country, many pragmatic approaches were proposed to explore the new world, just like babies do.

In 1984, Donald A. Schon published "The Reflective Practitioner: How Professionals Think in Action" (Basic Books) [5, 6].

Chapter 9 describes the meaning of pragmatic experiment or "Pragmatism" how it is increasing its importance in terms of strategic design.

## References

1. https://en.wikipedia.org/wiki/Walter_A._Shewhart
2. https://en.wikipedia.org/wiki/Charles_Sanders_Peirce
3. https://plato.stanford.edu/entries/abduction/
4. https://plato.stanford.edu/entries/abduction/peirce.html
5. https://en.wikipedia.org/wiki/Donald_Schon
6. D.A. Schon, *The Reflective Practitioner: How Professionals Think in Action* (Basic Books, New York, 1984)

# Chapter 8
# Pattern-Based Approach (PBA) to Learning from Failures

## 8.1 Patterns and Learning from Failures: Their Increasing Importance

When you talk to someone, you observe the listener's face, behavior, etc., and presume what he or she is thinking or feeling in his or her mind. You imagine whether your talk is truly accepted or agreed or not. Such presumption is the essence of communication. Communication is carried out without words. It is carried out by sensing multisensory or nonverbal information. Yes, communication is nonverbal. It looks as if we are conversing, but conversation is carried out in words or in signals. We don't need any presumption.

Patterns are one of these tools to represent nonverbal or multisensory information. We understand the situation at once. As described later, cartoon characters convey their emotion using patterns. No words are necessary. When we look at a character, we immediately understand his or her emotion.

Digital Transformation (DX) is attracting wide attention these days. But we should remember that "Analog" contains the whole information. DX is a selection of necessary data from the analog world. DX is a discrete world, but "Analog" is a continuous world. Most materials and living things are continuous. We need analog information to understand the whole.

Pattern is so to speak "Analog Transformation (AX)." We sense multiple information (note this is not data. It is Information). This is very much different from current AI, which is totally based on digital world. I believe the next AI will focus more on "Analog Information." How we can edit digital or separate information into analog or meaningful information will be the challenge of the next AI.

Thus, I firmly believe patterns will play a very important role in the new world Internet of things (IoT) will bring. It is another IoT, "Integration of Things" and Humans and Machines work together as team members *on the same team*. This quickly expanding world will call for integrations of things. There will be no walls between disciplines, business, or engineering sectors. Material Science and Life Science will be merged into one.



S. Fukuda, *Self Engineering*, SpringerBriefs in Applied Sciences and Technology,
https://doi.org/10.1007/978-3-030-26725-4_8

**Table 8.1** Five sectors of the economy

| Sector | Activities |
|---|---|
| 5 Quinary sector | Decision making |
| 4 Quaternary sector | Knowledge and ICT industry |
| 3 Tertiary sector | Service industry |
| 2 Secondary sector | Transforms raw material into products—manufacturing, etc. |
| 1 Primary sector | Extracts raw materials from nature—agriculture, fishing, etc. |

In addition to facilitating perception, patterns will play an important role in our decision. We act based on our situational perception. Patterns are tools to facilitate human decision making. Most of the decision needs to be done in a very limited time and with limited resources.

Economist is now proposing quinary sector. This is decision making. They insist the next economy will be decision making (Table 8.1). Engineers stick to problem solving. Very few discussions are made about decision making or what goal we should pursue. "How" is dominant in engineering and "What" or "Why" has been, to exaggerate a bit, ignored. Indeed, engineering education at the level of K-12 emphasizes the importance of "Why," but regrettably enough, engineering in industry is focused on "How."

Looking back, engineering started to make our dreams come true. When we are dreaming and trying to make it come true, we examined whether it is really feasible with own "Self" capabilities and with available resources. If it looks very difficult, we change the goal and pursue another dream. Thus, making dreams come true is a repetition of trials and errors.

We need to remind ourselves that engineering is our efforts to realize our dreams or to make our dreams a reality is engineering. Thus, patterns are expected to provide a good tool for deciding where we should head for to make our dreams come true. And as this is creation of a new world, we need to go ahead by trial and error. Learning from failures, therefore, is crucially important to make a step forward.

## 8.2 Performance Indicator

As situations change unpredictably and the world is expanding rapidly, you cannot do business on the "Tomorrow is another today" principle. Not only market is expanding, but you must tackle with a wide variety of markets. That is why economists pointed out the quinary sector of the economy is decision making. But to make a decision, we need a business performance indicator. In business, money constitutes the basic unit. So as far as quantification is concerned, there is no problem. The problem is "How" you convert value to money. With the progress off diversification and personalization, the conversion of value to money becomes increasingly difficult.

Therefore, "Key Performance Indicator" is getting wide attention these days in business. Identification of key performance indicator becomes necessary to develop business strategy. Trial and error was also inevitable in business, yesterday. But, at that time, situations did not change appreciably, or changed smoothly, so they could make errors within allowable limits and make another try.

But today, if we do not have a key performance indicator, it should be carried out in an organized way. Otherwise, they cannot determine their strategy (Fig. 8.1). Learning from failures will lead us to organized trial and error so that we can gradually establish the allowable limits and make adequate decisions (Fig. 8.2).

To stabilize the trials and reduce the errors and to keep performance within the permissible range, it is inevitable to identify the key performance indicator.

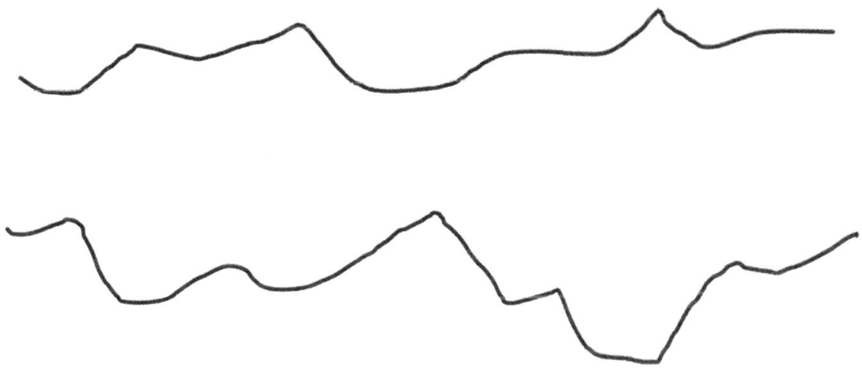

**Fig. 8.1** Unorganized trial and error

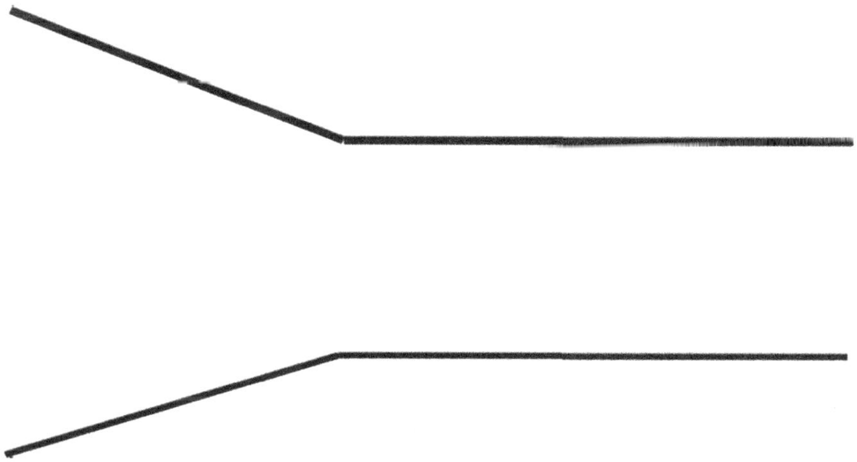

**Fig. 8.2** Organized trial and error

In business, identifying a key performance indicator is something like how we can prepare a dashboard which enables the viewer to understand the situation easily and help him or her to develop a strategy without too much difficulty.

In engineering yesterday, the model-based approach worked well, not only because the situations were predicable, but performance of each machine element can be measured. But in engineering today, machines are working together as a team and situations are becoming unpredictable, so finding a key performance indicator becomes extremely difficult.

But to reduce the trial range and keep the errors within the allowable range, we need to find out a performance indicator to control human movement, for example. Human motion trajectories vary widely from time to time, from person to person. So, it is not easy to find out a key performance indicator which works for all cases.

Many measuring instruments are developed to measure human motion. But they only show us the successful motion. We cannot reproduce this successful trajectory, because human motion varies from person to person and from time to time. And the current situation might be different from those at the time of measuring.

What we need to learn to control *human* motion is how we learn from failures and reduce errors. In other words, we need a quantitative index to understand whether we are doing better or worse.

And what makes learning of controlling human motion difficult is human body parts are closely connected and they work together in coordinated and balanced manner. This is a big challenge.

Further, only Somatic Nervous System relating to Kinesthetics is discussed often [1]. But Peripheral Nervous System [2], Autonomic Nervous System [3], which is composed of Sympathetic Nervous System [4] and Parasympathetic Nervous System [5], and Central Nervous System [6] all these work together harmoniously so that we can coordinate and balance our body motion. Especially Sympathetic Nervous System relates deeply to activate human bodies.

Timothy pointed out in his book "The Inner Game of Tennis" [7] that we must learn to recover from our failures by ourselves. The best tennis coach only says, "Good shot," but he does not say anything how the learner should recover from failures. So, a learner asked him "Why don't you teach us how to recover from failures?". His answer was "Muscles are different from person to person. So, you should learn by yourself."

In the case of tennis, however, there are rules and the outer world is fixed or explicit. But in the case of swimming, there are no rules and the outside world cannot be described verbally. It is a typical world of tacit dimension of Michael Polanyi [8].

Therefore, self-learning is the only way to learn to swim. But to do that, we need to know how much better we are doing this time than before. We need a measure to compare. But regrettably very few, if any, tools are provided for this purpose.

Further, we must note that most cases which are called human error are not an error at all. Humans did not follow the instructions, because they needed to move their bodies to adapt to the situation. So, they did not or to be more exact and could not move as they were expected. It is only because their motions exceeded the bounds of the frame the design engineers expect them to stay in. Therefore, it becomes increasingly important how we can establish the way to learn to control *human* motion.

## 8.3 Mahalanobis Distance (MD)

Mahalanobis Distance (MD) is a measure of the distance between a point *P* and a distribution *D*, introduced by P. C. Mahalanobis in 1936. It is a multidimensional generalization of the idea of measuring how many standard deviations away *P* is from the mean of *D* [9–11] (Fig. 8.3).

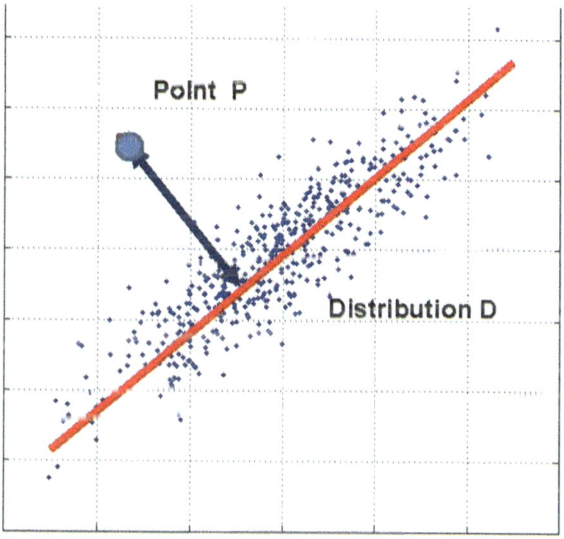


**Fig. 8.3** Mahalanobis distance (MD)

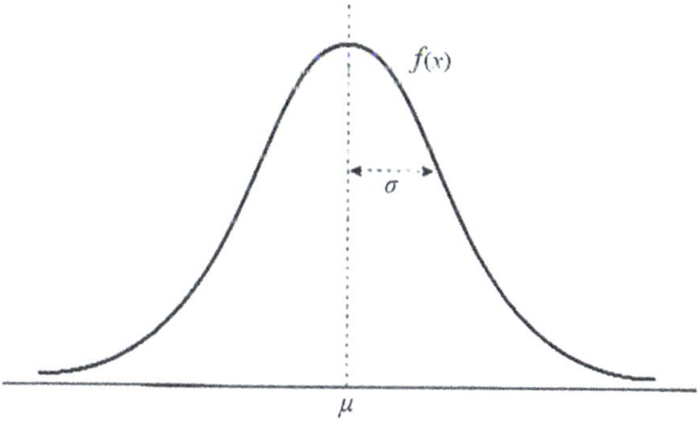


**Fig. 8.4** Normal distribution with mean $\mu$ and standard deviation $\sigma$

The normal distribution (Fig. 8.4) is the most important probability distribution in statistics because it fits many cases. But apart from rigorous statistical discussion, we know that if we repeat the process many times, the result comes out as a normal distribution.

MD is unique because it normalizes distance from standard deviation by the mean, i.e., $\sigma/\mu$. So, it is unitless and can be extended to multidimension without considering orthonormality.

Mahalanobis proposed MD to find out the outliers, because the outliers cause serious problems in statistical analysis. He was interested in principal component analysis (PCA). PCA can be described as follows [12].

Principal component analysis (PCA) is a statistical procedure that uses an orthogonal transformation to convert a set of observations of possibly correlated variables (entities each of which takes on various numerical values) into a set of values of linearly uncorrelated variables called principal components. Matthew E. Clapham explains PCA by video [13].

Mahalanobis would like to remove the outliers, because outlines affect the dataset, so that principal component may not be correctly obtained (Fig. 8.5).

Blue is correct principal component. But the point $P$ in red and other points (outliers) affect its distribution $D$ in blue and change the distribution to red and produce incorrect principal component (in red). Therefore, such outliers should be removed.

Further, principal components have orthogonal relations with each other (Fig. 8.6). As PCA is based on orthonormality, Euclidean distance can be used. To rigorously discuss PCA, Euclidean space and Euclidean distance are required. But units come with them. In fact, the purpose of multivariate analysis is to represent multivariate data in fewer dimensions. And eigenvectors describe directions in space.

But if the main purpose is to remove outliers, we do not need to stick to Euclidean distance. We can introduce a much simpler approach. That is MD.

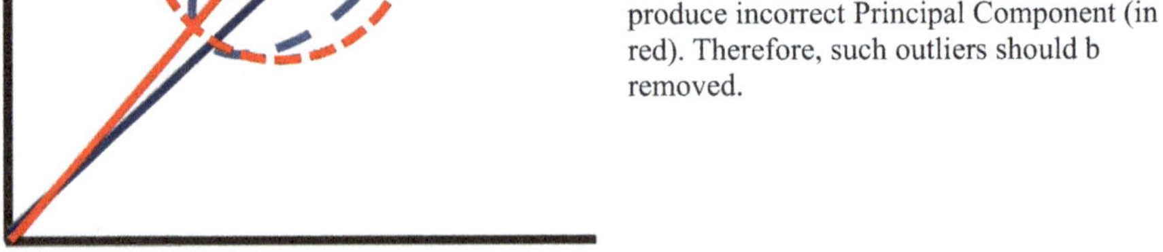


**Fig. 8.5** Outlier affecting principal component analysis (PCA)

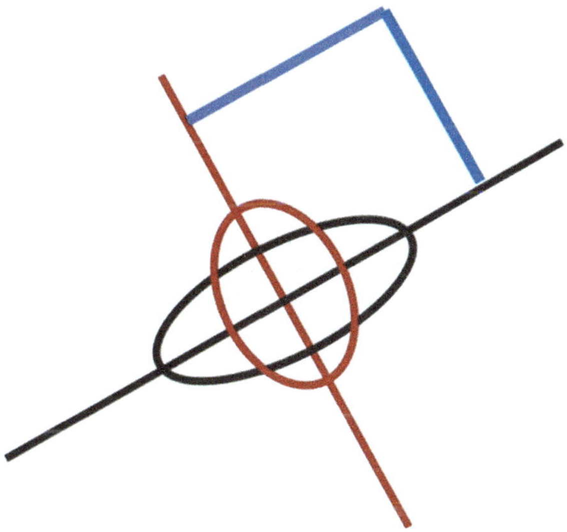

**Fig. 8.6** Principal components are orthogonal with each other

What matters is how much far away the point is from other observed data. If MDs of such points are too large, it means that they are far away from other observed data. So, possibly they are outliers. They may be there due to some errors or due to some other reasons, and it is not appropriate to consider them as good and reliable data.

Just to remove the outliers from good dataset, it is enough to know deviation/mean ($\sigma/\mu$). If this is too large, it means that the point is too far away from other data. So, it can be regarded as an outlier and can be removed. This is very simple and does not call for any orthogonality. Thus, it can be easily extended to multidimensional applications, and further, we do not have to care about the unit. Thus, this multidimensional unitless metric approach can be established.

Although the original aim of Mahalanobis is just to simplify the procedures of finding out outliers, MD is coming to be used more and more today, because in addition to its simplicity, due to its unitless feature and no constraint of orthogonality, it can be used for many different cases where each sample data are composed of many different units.

## 8.4 Mahalanobis–Taguchi System (MTS)

Taguchi realized that although Mahalanobis was interested in finding outliers and to exclude them to secure good dataset, he can utilize MD in a different way, if he introduces patterns. Then, Taguchi quality, which is different from the regular definition, can be controlled very easily.

This book explains why the pattern approach is useful for learning from failures by taking controlling human motion as an example. Therefore, we will not go into detailed discussion of Taguchi Method [14–16]. We will describe it as much simple as possible, although the discussion is too much rough from statistics point of view.

What differentiates the definition of quality between Taguchi and most other researchers is Taguchi considers dynamic control of quality. He pointed that the current quality control is insufficient for dynamic control. He asked, "How do you define variance of the hands of the clock?" Usual statistics discussion is about the static data. They may be considering the dynamic functions, but the data themselves are static. But Taguchi's primary goal is to control the quality dynamically or to control the quality from temporal perspective. Others are discussing how to control the quality in space perspective or in perspective at a certain time.

One of the points which distinguishes Taguchi's definition from most other ones is his quality is based on Loss. Usually when we discuss quality, we are thinking how better functions can be added. But Taguchi's definition is substantially the same as Cost Performance Index (CPI), i.e., how we can reduce cost. He introduced the idea of Loss Function [17, 18] to reduce the cost of flow in production.

Another point characterizing his Loss Function is he considers the Loss of the Society. Industries emphasize the importance of cost reduction. But if the reduction is made at the expense of social loss, then the company cannot continue their way or lose money in the long run.

So, he emphasizes the importance of considering loss from the society perspective. In other words, his Loss Function is how to minimize the energy consumption and it truly matches the concept of sustainability.

Taguchi is very well known for his Taguchi Method. Taguchi Method uses orthogonal table, and it is orthonormality-based. He introduced his original S/N ratio. This is very much different from S/N ratio used elsewhere. To describe Taguchi's S/N ratio in a simple way, it is the distance of *P* from the distribution of Noise. Thus, roughly speaking, the basic idea of variance/mean is the same as MD. But Mahalanobis used it for removing outliers, but Taguchi used MD for controlling quality. If MD is large, we need to control the parameter to stay within the acceptance level. If MD is within the acceptance level, then we do not need to pay any special attention to each parameter.

To describe this in another way, if the parameters are represented as patterns, and the pattern is within the acceptance level, i.e., the pattern may be regarded as identical, then his quality is controlled successfully. If the pattern cannot be regarded as identical, then we need to check how and why the pattern is different from the acceptable pattern, and some measures must be taken.

Figure 8.7 illustrates the idea. Unit space is created based on typical samples (Fig. 8.8). And MD is calculated for each quality pattern. If MD of the pattern exceeds the threshold value, then that pattern is not acceptable, i.e., quality is not good. We should note that we use pixels to produce a pattern (Fig. 8.9), so that any images can be represented as patterns.

This idea was very much welcomed by industries which cannot control quality element by element. What is better, this approach does not call for orthogonal arrays. So, it is quite easy for them to control quality not on an element basis but on the whole product or production basis.

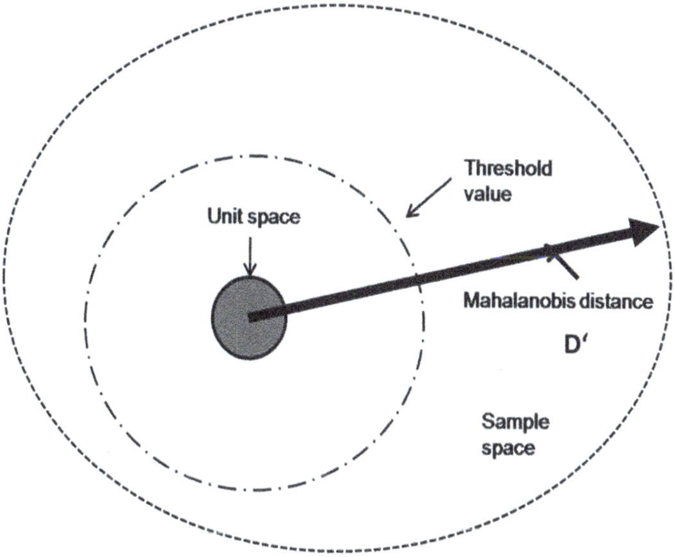


**Fig. 8.7** MD-based pattern identification

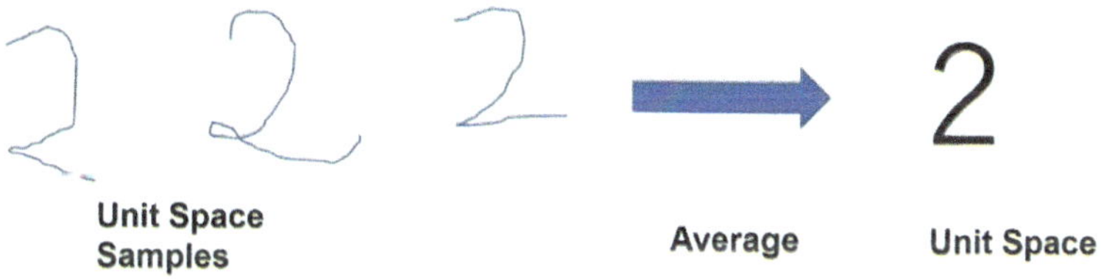


**Fig. 8.8** Creating unit space

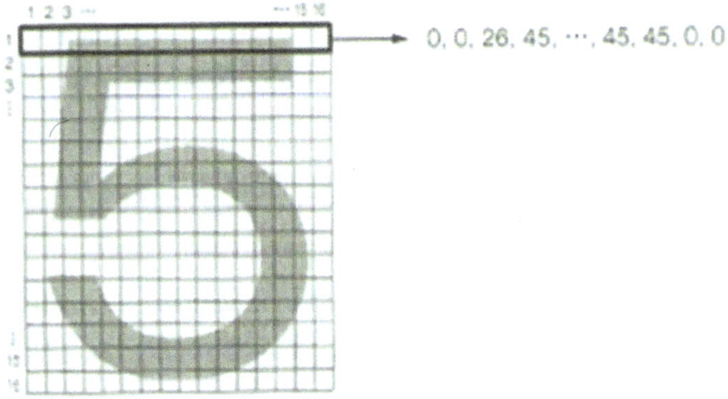


**Fig. 8.9** Representing as a pattern

Thus, Mahalanobis–Taguchi System (MTS) [19] spread very widely in a very short time. So, MTS is now being applied in many different ways in many different fields.

But some industries would like to control quality element by element. So, for them, Mahalanobis–Taguchi–Schmitt Approach was developed [20], which utilizes orthogonal arrays. Those people are against the approach without using orthogonal arrays. They think the dataset might include the outliers so that the dataset is not properly represented. But if we cannot carry out design of experiments which is element-based, then what else can we do? We have to put up with the dataset at hand. We have no other way.

Let me repeat once again. We must remember that Taguchi's definition of quality is totally different from the usual definition of quality. When we say quality, we think of how we can have better working products. But Taguchi's definition is how we can reduce the loss in production. So, he introduced Loss Function to evaluate quality. When the Loss Function is reduced to the minimum (Fig. 8.10), then quality is secured.

This may sound very strange to quality control engineers, but this is very popular in industry. Industry emphasizes cost performance. Taguchi's assertion is, in other words, to reduce the cost as much as possible. What differentiates his assertion from industries is he is not thinking industry alone. He is thinking about society. Even If one industry succeeds in achieving the best cost performance for that particular company, but if it introduces large cost to the society, then that strategy would not last long and from the social point of view, it is not controlling quality in the appropriate way. The loss should be minimized from the social point of view. This is his assertion. Thus, pattern approach using MD is widely introduced in many fields, not only in quality control, but in other fields as well.

Let us consider Taguchi's Approach of Loss Function from another perspective. It is not too much to say that Taguchi knows human psychology very well. Of course, everybody would like to have better products, but once he or she has it, he or she would like to have one more step better product. So, they continue wanting better products. But of course, that requires a lot of money and efforts, but such investments are not always successful. If the customers are not satisfied, then what a large damage that industry will suffer. But even if your product remains the same in terms of functions, etc., you will feel satisfied, if the quality remains the same, no matter under what situations you are using them.

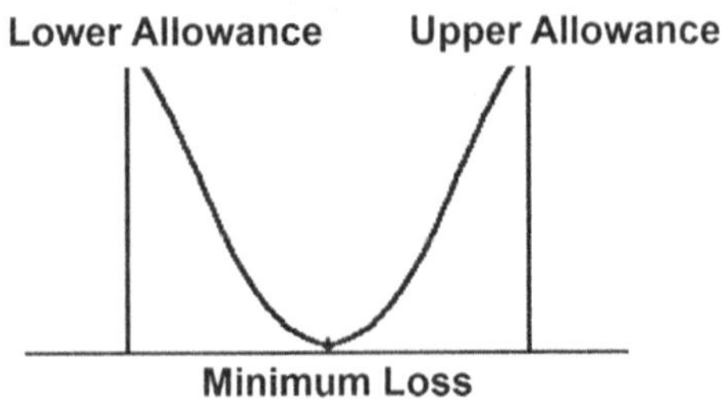


**Fig. 8.10** Loss Function

It is not a homerun, but a small hit. But if the player strikes a hit every time, even though it may be small, then you can trust that player. This is the problem of trust, and it is nothing other than establishing brand for the company.

As the idea of Loss Function includes Loss at the Society level, we cannot introduce orthogonal arrays easily. We cannot carry out Design of Experiments with the Society included. The problem is too much complicated and complex. Thus, the idea of the Loss Functions serves to secure trust, and it leads to establishing brand. Taguchi's pattern-based MD approach is widely accepted by a large number of industries.

## 8.5 Mahalanobis–Taguchi–Fukuda Approach (MTF)

The uniqueness and usefulness of Mahalanobis–Taguchi System (MTS) is to define a quality as a pattern. In the conventional quality control, individual quality elements are controlled to achieve the required quality of a final product. But MTS introduced a pattern approach so that industries can control quality without paying attention to individual elements.

MD does not require orthonormality, although Euclidean distance does. The primary purpose of developing MD is to remove the outliers, so the focus is to find out whether the point is an outlier or not. Mahalanobis did not care about the whole. His interest was individual cases. But if he uses Euclidean distance, he must consider the whole. So, MD is very good for individual cases to find out outliers that should be removed, but at the time of Mahalanobis, it was considered that MD does not work well to grasp the whole.

Taguchi defied this conventional idea. In fact, he is a challenger and challenges many conventional ideas, as S/N ratio, his Quality definition, etc. show. He demonstrated if we introduce a pattern approach, MD can be utilized for grasping the whole. And better still, we can ignore units. This is a great contribution of Taguchi. Therefore, MTS is accepted widely in industry. Indeed, in some industrial fields, it is impossible to control quality element by element.

Taguchi developed this pattern approach, primary because his S/N (Signal/Noise) ratio idea is basically identical with Mahalanobis Distance, which is $\sigma/\mu$ (=Standard Deviation/Mean), his definition of function is an input–output relation, and it is not just a characteristic value in other definitions of quality control engineering.

MTS reminded Fukuda of the importance of usefulness of a pattern approach. One of his research interests is Emotional Engineering [21]. He used to study detection of emotion from face. His group tried many different image processing techniques, but they are too much complicated and take a lot of time. Further, the results were not satisfactory. Then, he realized that we can understand emotions of the characters in cartoons easily. At that time, cartoons were in black and white, and characters were very simple. But we can understand their emotions at once. So, he introduced the cartoon face approach and his group succeeded in detecting emotion from face easily in a very short time (Fig. 8.11), [22, 23].

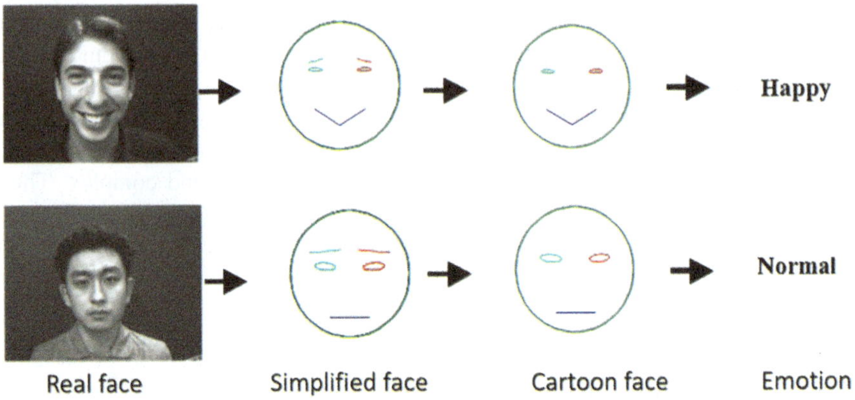


**Fig. 8.11** Cartoon face model

At that time, his team used cluster analysis to group datasets. And in this series of research, they found out that acceleration plays an important role for emotion detection. Thus, they realized the importance of dynamic approach.

When Fukuda came across Taguchi's pattern approach using MD, he realized that as MD and cluster analysis are basically identical, he can introduce Taguchi's pattern approach to motion control. Taguchi was interested only in static pattern identification, because his interest is quality control. But Fukuda's interest is movement. So, the problem here is dynamic. Therefore, Fukuda extended Taguchi's approach to dynamic pattern recognition.

Fukuda noted these features of MTS and introduced Taguchi's Pattern Approach to human movement control due to the following reasons.

As Nikolai Bernstein pointed out [24], the biggest problem of human movement control is its abundantly large number of degrees of freedom. Human movement trajectories vary widely from time to time (Fig. 8.12).

Bernstein's discussion is with respect to one person's movement. What makes the problem more difficult is human movements vary from person to person. So, the traditional engineering approach of system identification and identifying parameters does not work.

Further, in the case of swimming, we cannot predict water behavior at all. So outside and inside, we have no ways to apply explicit or rational approaches. Swimming is a typical example of tacit dimension. Therefore, we need to learn by trial and error.

Why our movements vary from case to case at its early stages is considered to be because we coordinate and balance our body parts to stabilize our movement by trial and error. In fact, when we get close to the target, our movements are stabilized, and our trajectories are fixed. Koike [9] points out that our muscles harden at this stage. Thus, we reduce the number of degrees of freedom to the minimum. Most of the motion control research focuses their attention to this stage and machine motion control is considered to reproduce this trajectory.

**Fig. 8.12** Cyclogram of hammering down

Interestingly enough, when we discuss controlling human movements from outside, we call it "Motion Control." But when we discuss it from inside, we call it "Motor Control." "Motion Control" developed from mechanics. So, it started from rational or explicit world. But "Motor Control" comes from medical and physiological fields. "Motor Control" is a very difficult issue, and most problems still remain unsolved.

We know very well that we have five senses. Sensory processing affects motor skills a great deal. But senses associated with motor control are more. The sixth sense, the vestibular system (simply put, ears), controls balance and spatial orientation. Proprioception is often called the seventh sense, and it senses the relative position of body parts and strength of effort to control movement. To describe it very roughly, it controls our musculoskeletal system and joints. The brain integrates these two pieces of information, but how they are integrated is still not clear.

Thus, sense of balance or equilibrioception plays an important role in our motion or motor control. Further, there are interoception and exteroception. Interoception is the sense of the internal state of the body, and it relates to emotion. And exteroception is the responsiveness to external stimuli. Thus, there are so many senses, and they work together in a very complicated and complex manner. Therefore, we do not know anything about their mechanism.

Therefore, we understand easily why we cannot follow the successful motions and control our motion in the same way. Many measurement tools and equipment have been developed to record our movements. But they show us only successful cases. But we cannot follow such successful movements, because our bodies and our ways of coordinating and balancing our body parts vary very widely from person to person. We are humans, not machines. We have no other choice than to learn by ourselves by trial and error.

Although there are many senses and they work together in a complex and complicated manner, Taguchi's pattern approach using MD is the best available tool for learning movement control. Its greatest benefit is learners can understand how much better or worse they are doing this time. They can evaluate the improvement or deterioration in a quantitative manner. They cannot tell how to improve, because this is related to the sense of balance and coordination, but they can *feel within their body* how they can improve.

Thus, MTF's approach is pattern-based, but it should be emphasized the final decision is left to us, humans. This approach is not to automate the process, but to prepare information for a human to judge and to decide. MTF is a support tool for decision making.

Fukuda has been studying Emotional Engineering for a long time [21]. His definition of Emotion is the following.

Emotion constitutes "Perception → Motivation → Decision Making → Action → Emotion" cycle, and we move and act by repeating this cycle.

This definition comes from the fact that emotion and motivation come from the same Latin word "movere." Thus, emotion means e=ex=out, so it means to move out.

Interoception is related to emotion. So, if we wish to control our movements, we need to consider not only kinematics, but also mental side within ourselves.

Move your body and record your movement, and measure MD. Then MD gives you the measure whether your movement is appropriate or not. If MD is within the threshold, your movement is fine, no matter in what way you move your body. But if MD is greater than the threshold, that movement is not appropriate. Keep on and if MD increases, then you are moving your body in the wrong way. Your movement is deteriorating. If MD decreases, then it means you are improving your body movement. So, keep on and then you will gradually learn to move your body in an appropriate way.

Patterns can be selected based on musculoskeletal systems or in the case of swimming, swimming suits or wearable robots with appropriate sensors can be used. Again, we must remember even images can be patterned if we represent data in pixels.

Thus, we can accommodate coordination and balancing of our body movement, which mechanisms are still unknown. Mahalanobis–Taguchi–Fukuda (MTF) Approach provides a quantitative guideline for the learners to learn how to control their body movements by trial and error.

Why we use MD for controlling human movements is because in Euclidean distance, all units must be the same. For example, if we compare only height, then we can use meter or feet for all data. But why human motions vary so widely is because our bodies are built very differently from person to person. Some are slender. Others are fat. Some are tall. Others are small. So how we move our bodies vary depending our body build. But to characterize how the learner's body is built, we need to use many different units, as meters, kg, etc. As to the mean and covariance of each, we can measure because they are in the same unit. MD transforms the random vector into zero mean vector with an identify matrix for covariance. Thus, we can utilize MD for all different kinds of body build, and the learner can use MD for measuring the improvement of his or her own learning. If you are interested in studying how MD is different from Euclidean distance, visit [9, 25] for example.

Another important point is that MD enables Unsupervised Learning, i.e., learning without a teacher, Unsupervised Learning is associated with Hebbian Theory, the theory to study the brain neurons during the learning process. Unsupervised Learning is called "Self"-Organization in Social Sciences [26–28].

In the case of swimming, we have no other choice than to learn how to swim by ourselves. No one can teach us how. So, MD is a very good tool for learning how to swim.

## 8.6 Applications

Before going on to the discussion about applications, let us look at the history of engineering development again (Fig. 8.13). Engineering (after the Industrial Revolution) focused on individual products for a long time. Our life sphere was small, and it was a closed world with boundaries. And changes did no take place appreciably and if they did, they were smooth, so we could predict the future. Engineers could foresee the operating conditions. So, functions, their reproducibility, and their robustness were most important. Indeed, it was an explicit and verbal world. Although the world expanded with the progress of engineering and it changed to an open world without boundaries, engineers developed "System Identification" approach and succeeded in establishing a controllable world, which is the extension of the rational

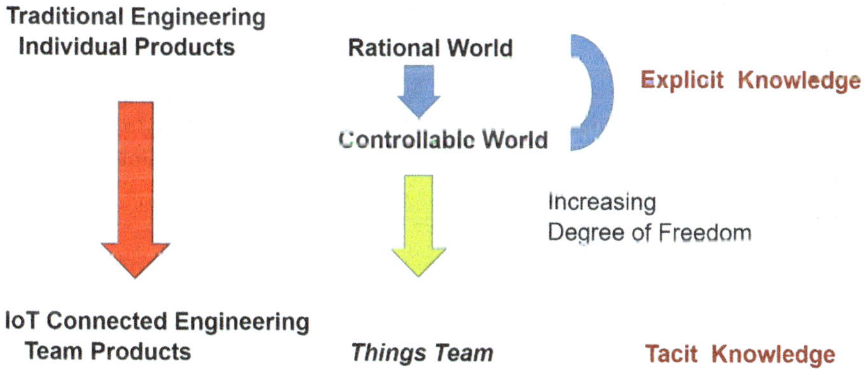


**Fig. 8.13** From traditional engineering to IoT-connected engineering

world. This success accelerated engineering progress and expansion of our life space tremendously. Accordingly, degrees of freedom increased exponentially along with them.

And changes are changing. They come to change in an unpredictable manner, so we must adapt to these changes which take place frequently and extensively in an unpredictable manner. Thus, "Adaptability" becomes most important. In other words, processes are becoming more important than products. To cope with such complex and complicated situations, we, humans, need to form a team and machines are also called for to work as a team. Thus, team working becomes a very important issue. Why we need team working is because we can cover more areas with less overlook.

Rapidly progressing Artificial Intelligence (AI), especially Deep-Learning, helps us to reduce the overlook and to cover very wide areas (Fig. 8.13).

But soon, IoT, Internet of things, was proposed, and humans and machines come to work together as team members *on the same team*. And as "Internet" indicates, "Communication" becomes important. Until then, what we have been doing was conversation, just an exchange of messages or signals (Fig. 8.14).

As the world becomes wide open without any boundaries, and it changed from explicit and verbal world to tacit and nonverbal world, we need to perceive multi-sensory information to make an adequate decision for the next action.

Thus, team formation and management become crucially important. The wall between engineering and business worlds is gone. Both have the same problem. Table 8.2 shows the type of teams and the current status of research and tools. Sport-type team is a typical example of the case where the number of team members is fixed, there are strict rules, and the goal is clearly defined.

So this is the game of perfect information. Therefore, the current AI will work very well. And most teams in the current society belong to this type.

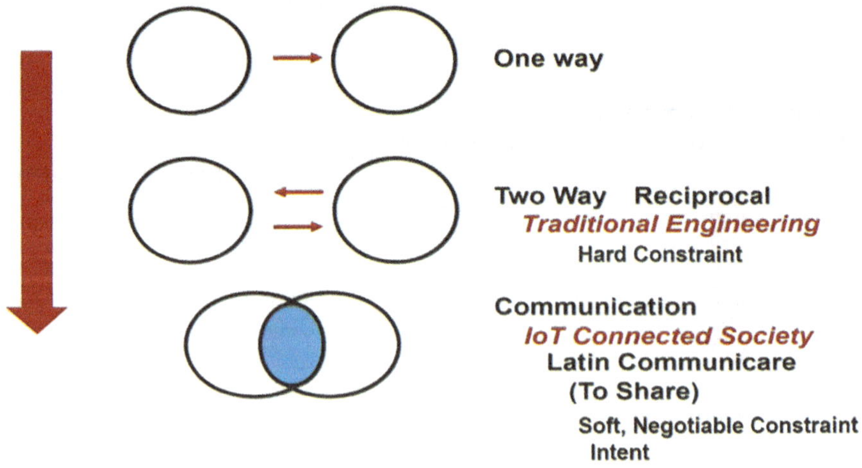


**Fig. 8.14** From conversation to communication

**Table 8.2** Team organization and management

| Team | Amount of research | Tool |
|---|---|---|
| Sports type | Enormous | Current AI |
| Adaptive team | Still few | Next-generation AI |
| Adaptive "*Things*" team | None (to my knowledge)<br>IoT-connected society | None |

But as the complexity and complicatedness increase in our world, we need to form an adaptive team. But in most cases, current adaptive team is Rockne's Best 11 team. Each member has diverse capabilities and can tie up with other members in response to the changes so that the team can adapt to the strategic change flexibly. In short, the current adaptive team is a sport-type team with increased adaptability. The true adaptive team is the team formed without any constraints. No constraint about the number of team members, its formation, etc. In other words, it is a network changing flexibly to adapt to the change of the situations and environments. Research on such network formation is still very few. But this discussion is only about human teams.

IoT will lead us to the world where humans and machines work together as team members *on the same team*. We have never experienced such a world, and it is indeed creation of a new world. Thus, we have no other choice but to go ahead by trial and error. Thus, learning from failures will become very important.

But here we must remember that as this is exploration or creation, we need to understand the environment and situation correctly to step forward adequately. In other words, self-learning is important. This self means not "Self." It means you have to learn by yourself.

Come to think, humans coordinate and balance many body parts to move. We rely on self-learning to make movement. We do not know the mechanism, but we know how we can improve our movements to do the next action. In engineering, we have been trying to make everything verbal and rational. But we should remember there is another engineering, which make the most of human capabilities. We have been ignoring human capabilities up to now and regarded humans as another machine element. But humans can judge and make decisions. This differentiates humans from machines.

Better still, humans do not cost too much in terms of money, energy, etc. Humans are born to be friendly with nature. So, we should consider to make use of humans more. But we should not forget to give them satisfaction. Self-actualization is self-motivated and by achieving self-actualization, we feel very much happy and satisfied. Then, we should change engineering from the current "Do it for someone else" and "Expert in each field" framework to "Do it yourself" framework. Although experts may still be needed, most people can live on their own and enjoy their life. This will certainly contribute to the Sustainable Development Goals (SDGs).

The originality of Mahalanobis–Taguchi–Fukuda approach is to rely on the human capabilities for making adequate judgment and decision making against the current engineering approach of automating everything.

Thus, the approach described here is expected to work widely in different fields. For example, it can work for developing wearable robots to help humans to move in his or her own way. Doing sports or jobs our own way reduces fatigue and increases joy.

The basic idea of this book is "We should make more use of human capabilities." This approach is expected to work in many fields. Needless to say, IoT is one of such fields.

And finally, to make most use of human capabilities, we should do our best to execute Analog Transformation (AX) no less than DX (Digital Transformation). It is important to know that even if the brain stops its neural activities, the other parts of our body keep running for about 3 h, if we keep blood flowing through the brain. This fact indicates that we should consider analog and digital world at the same time.

In other words, we should consider the issue in a holistic way. Then, we can get to the goal, no matter what path we may take.

## References


1. https://en.wikipedia.org/wiki/Somatic_nervous_system
2. https://en.wikipedia.org/wiki/Peripheral_nervous_system
3. https://en.wikipedia.org/wiki/Autonomic_nervous_system
4. https://en.wikipedia.org/wiki/Sympathetic_nervous_system
5. https://en.wikipedia.org/wiki/Parasympathetic_nervous_system
6. https://en.wikipedia.org/wiki/Central_nervous_system
7. W. Timothy Gallwey, *The Inner Game of Tennis: The Classic Guide to the Mental Side of Peak Performance*. (Random House, New York, 1997)
8. M. Polanyi, *The Tacit Dimension* (University of Chicago Press, Chicago, 1966)
9. https://en.wikipedia.org/wiki/Mahalanobis_distance
10. https://www.youtube.com/watch?v=spNpfmWZBmg Multivariate Statistics Mahalanobis Distance by Matthew E. Clapham
11. https://www.youtube.com/watch?v=3IdvoI8O9hU Intuitive Understanding of Mahalanobis Distance by Gopal Malakar
12. https://en.wikipedia.org/wiki/Principal_component_analysis
13. https://www.youtube.com/watch?v=TSYL-oHx4T0&t=561s 28 Mathew Ordination Method Principal Component Analysis by Matthew E. Clapham
14. https://en.wikipedia.org/wiki/Taguchi_methods (There are many books about Taguchi Method. But for our purpose, this is enough. This contains a little bit of criticism from the regular statistics standpoint as well)
15. http://www.ecs.umass.edu/mie/labs/mda/fea/sankar/chap2.html
16. https://www.ee.iitb.ac.in/~apte/CV_PRA_TAGUCHI_INTRO.htm
17. E.A. Cudney, K. Ragsdell, K. Paryani, quality loss function: a common methodology in three cases. J. Ind. Syst. Eng. **1**(3), 218–234 (2007) (There are many books about Taguchi's Loss Function. But when you retrieve, be sure to indicate Taguchi Loss Function. Otherwise, other loss function come up, too)
18. P.J. Ross, *Taguchi Techniques for Quality Engineering: Loss Function, Orthogonal Experiments, Parameter and Tolerance Design*. (McGraw-Hill Professional, New York, 1995). (This book describes Taguchi's original ideas very well and covers them all. But we do not need to study this deep for our purpose)
19. G. Taguchi, S. Chowdhury, Y. Wu, *The Mahalanobis Taguchi System* (McGraw-Hill Professional, New York, 2000)

20. G. Taguchi, R. Jugulum, *The Mahalanobis-Taguchi Strategy: A Pattern Technology System* (Wiley, Hoboken, NJ, 2002)
21. S. Fukuda, *Emotional Engineering*, vol. 1–7. (Springer, London, 2011–2019)
22. V. Kostov, H. Yanagisawa, M. Johansson, S. Fukuda, Method for face-emotion retrieval using a cartoon emotional expression approach. JSME Int. J. **44**(2), 515–526 (2001)
23. V. Kostov, S. Fukuda, M. Johansson, Method for simple extraction of paralinguistic feature in human face, image & visual computing. J. Inst. Image Electron. Eng. Japan **30**(2), 111–125 (2001)
24. V. Kostov, S. Fukuda, Emotional coding of motion using electromagnetic 3D tracking instrument, applied electromagnetics, material science and device. Japan J. Appl. Electromagnet. Mech. **8**, 229–235 (2001)
25. https://www.quora.com/What-exactly-is-Mahalanobis-distance-How-is-it-different-from-Euclidean-distance
26. http://mccormickml.com/2014/07/22/mahalanobis-distance/
27. https://en.wikipedia.org/wiki/Covariance_and_correlation
28. https://towardsdatascience.com/importance-of-distance-metrics-in-machine-learning-modelling-e51395ffe60d

# Chapter 9
# Strategic Engineering

Most activities of engineering are tactics-focused or problem solving. The goal is clearly defined, and how the problem can be solved is pursued. If we call such activities in engineering, Tactics Engineering, there are other activities in engineering, which are strategy-focused or goal finding. In such Strategic Engineering, finding and achieving the goal are pursued. But such Strategic Engineering activities are quite a few. Indeed, team organization and management is one of such Strategic Engineering activities, but in most cases, the number of team members is fixed and there are rules. Team formation and management in sports are the goal clearly defined, and the number of players is fixed and there are strict rules. So, before a game, team formation becomes an issue, but once a game starts, how the team is managed is the only issue. In other words, the team formation or network does not change, although what is expected from each player may vary greatly from game to game. How effectively and flexibly the network can be managed to adapt to the changes becomes the problem. Thus, this issue can be solved by recurrent neural network. And such emerging technology as Reservoir Computing (RC) will contribute to its solving. It is demonstrated, for example, that RC can deal with the problem of gripping soft materials [1]. So, the current framework of AI is expected to be very much effective, if we ignore its huge energy consumption problem. In fact, Watson and AlphaGo consume 100–200 kW. Our brain, on the other hand, consumes only 20 W. Thus, it is soluble in principle, but in reality, this is a very difficult challenge.

In the case of human movement control, we call the movement control from outside, Motion Control and control from inside, Motor Control. Motion Control is based on mechanics, and machine movement is controlled by Motion Control. But most of the issues related to Motor Control are tacit and nonverbal, so that most of them still remain unsolved in spite of tremendous efforts of medical and physiological researchers. It is known, for example, that Proprioception plays an important role in coordinating and balancing our body parts, but its mechanism is not clear.



S. Fukuda, *Self Engineering*, SpringerBriefs in Applied Sciences and Technology,
https://doi.org/10.1007/978-3-030-26725-4_9

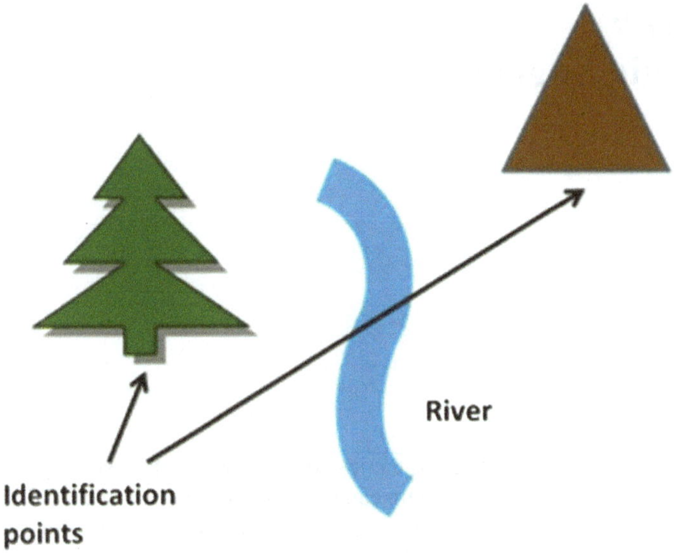


**Fig. 9.1** Identification of the name of a river

These are attempts to clarify the mechanism and to express it explicitly and verbally. But if we note that even if we may succeed in clarifying mechanism, the cost or energy needed might be huge as the case of Watson and AlphaGo demonstrated.

Further, we should take note that in these attempts, the system is identified. But if we consider swimming, we cannot identify the system at all. In fact, the idea of system identification is the same as identifying the name of a river (Fig. 9.1).

If we look at the river itself, we cannot identify its name. But if we look around and find trees or mountains that do not change, then we can identify its name. This is the basic idea of system identification, and once a system is identified, we can find parameters to control. Thus, although the world expanded from a small and closed world with boundaries into an open world without boundaries, we succeeded in expanding the rational world to the controllable world. System identification is the source of engineering development.

But in the case of swimming, we cannot identify the system. And such emerging new engineering framework as IoT poses the same problem. In such environments and situations, we need to find a solution by trial and error. What becomes important is whether we can achieve the goal or not and how we get there is no more a problem.

Changes occur very frequently and extensively today. But we must take note that changes took place yesterday, too. But changes are changing (Fig. 9.2).

Yesterday, changes were smooth, so we can differentiate them mathematically. In other words, we can predict the future. But today, the changes are sharp. So, we cannot differentiate them, i.e., we cannot predict the future. We are swimming in the flow of environment and situational changes.

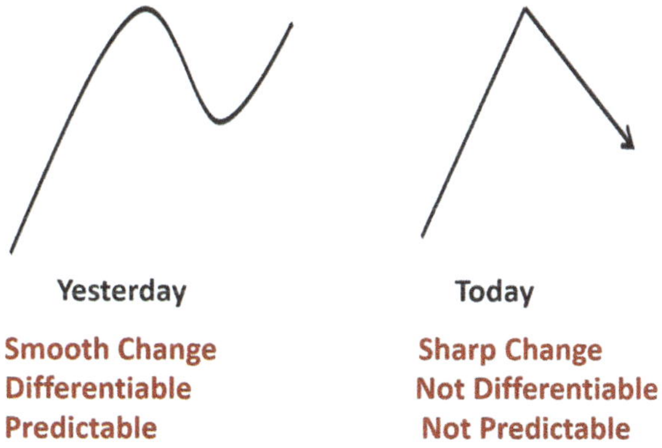


**Fig. 9.2** Changes of yesterday and today

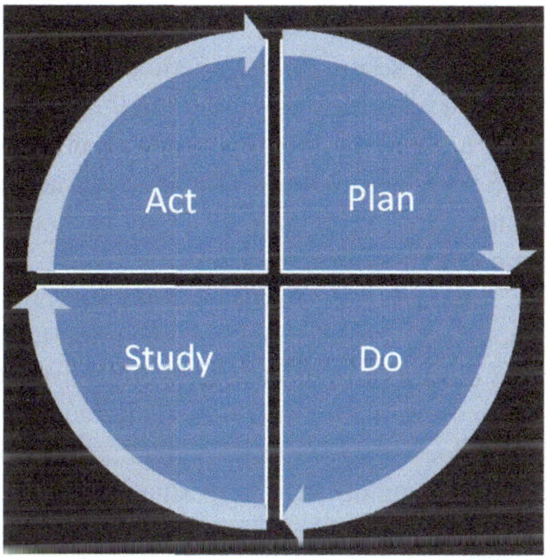


**Fig. 9.3** PDSA cycle

Then, how can we tackle with this problem? It is by trial and error, and we must learn from failures. Note that this failure means "fail" in the sense of "fail one's expectation." Not the fatal failure.

Many such pragmatic approaches have been developed. Peirce [2] proposed Abduction Theory [3, 4] in the years of Westward Movement. He is called the Father of Pragmatism. As this is truly exploration, nobody knows anything about the West So, "Abduct" hypothesis and apply it. Then, study if it works. If it works, then go ahead to explore. If not, look for another hypothesis. Repeat this cycle until working hypothesis is found. This is Abduction theory.

In fact, basically, this is the same idea as PDSA (Plan-Do-Study-Cylce). Both are based on Pragmatism. But Peirce is the father of Pragmatism. And Abduction Theory was based on his philosophy of Pragmatism, logics, while PDSA was developed by Walter Shewhart, based on his quality control approach.

Shewhart [5] proposed PDSA (Fig. 9.3) in 1939.

This clearly indicates how cycles are run. "Plan → Do → Study → Apply" cycle is repeated until the goal is achieved. He is known as statistical control expert. But his statistics is very much different from other statistical control approaches. These truly statistical approaches process data without any prior hypothesis. They analyze data using statistics to get a result.

In short, PDSA is a strategic approach, while statistical approach is tactical.

Shewhart's approach is Design of Experiments. He uses statistics to verify if his hypothesis works or not. In statistics, there are two approaches—Statistical Data-driven Approach and Design of Experiment Approach. The latter is goal-driven and pragmatic. Its basic is to study by trial and error.

The word "Design" comes up. We should remember that "Design" is creating a new world. Or to express it another way, it is exploration. But today, most of engineering design is problem solving. The goal is fixed, and engineers make efforts to how to develop the final product that meets design requirements.

"Design," however, is creation of a new world so goal finding is important. It does not matter so much what methods we use to get there. The fact we can arrive at our goal we set is most important. In this sense, PDSA truly represents what Pragmatism is. It is a way of thinking how we can explore the new world.

Schon [6] proposed "Reflective Practice" and published "The Reflective Practitioner: How Professionals Think in Action" in 1984 [7]. This also emphasizes the importance of Pragmatism.

Although many researchers pointed out the importance of Pragmatism, regrettably enough they pointed out only the importance of the reflective or cyclic approach, and they did not show us how we can improve the search or how we can learn from failures. We must do better next time. Otherwise, we will waste time. This is apparently due to the lack of appropriate performance indicators. Especially the fact that we do not have any quantitative performance indicator for trial and error is considered to be the reason.

The need and importance of such performance indicator has been strongly recognized in business, because strategy is important in business deployment. Business management is fundamentally exploring new markets. So, it is Design and Exploration. But in business, they have their own quantitative measure, i.e., money. So, they can make strategic decisions based on this quantitative performance indicator.

But in the case of human movement control, we do not have such performance indicators, needless to say, quantitative performance indicator. If we consider swimming, this can be easily understood. But to learn swimming, we need to know how much we are improving or deteriorating. Otherwise, we cannot learn.

In fact, learning by doing is emphasized very often, but there are no explanations how we can improve by doing. To achieve this, we need a quantitative performance measure.

Another hurdle to develop such quantitative performance indicator is Euclidean Space requires orthogonality and units come together. As the idea of system identification cannot be applied to water, we need non-Euclidean Space approach without any need for units.

Mahalanobis Distance (MD) solved this difficulty. It is a multidimensional generalization of how many standard deviations away the point $P$ is from the mean of the distribution. So it does not require any units and orthogonality. Taguchi proved how effective it is if quality is represented as patterns. In other words, Taguchi demonstrated that MD can effectively be applied to holistic recognition of quality. This, however, is static pattern identification.

Fukuda expanded a pattern-based approach and moved it two steps ahead. Swimming is dynamical so he expanded the approach to accommodate dynamic changes. He made one more step ahead. It is related to "Self." "Self" is emphasized as the most important need of humans. Why? It is because it is our attempts to expand the world or community of human species. Self-actualization is not personalization. We are driven to actualize ourselves in many different ways, and it is our human attempts to expand our community of human species. This is indeed nothing other than exploration of the new world. Once our minimum level of sustainability is secured, our needs go up and we would like to develop or it would be better to say, explore the new world to find the possibility of expanding our community of human species.

Self-actualization calls for decision making on our own. Decisions vary from person to person so diverse actions are taken. These actions interact with the outside world, and those that harmonize with it will remain and develop further. This is evolution. Self-actualization is our attempts to evolve.

Today, we are now in the midst of flowing changes. We need to change our engineering from tactics-focused to strategy-focused. The pattern approach introduced here as swimming as an example is expected to work in such environments and situations. IoT, for example, needs strategic engineering. As humans and machines work together on the same team, we need to remove the barrier between Material Science and Life Science. And we need to integrate many disciplines. So, strategy becomes crucially important.

Self-actualization is motivated by our needs as human species. So, we need to re-consider the importance of emotion and other elements of instinct. As the human represents vertebrates, our way of thinking is opt to be brain-centric. But if we look at the octopus, which represent invertebrates, it interacts with the outside world directly. As the octopus parents die soon after their babies are born, the octopus do not transfer knowledge from generation to generation. They have to live on their instinct alone. And in the case of the octopus, they are called Cephalopod (head + foot) and there is no body. But they can survive any environments and situations.

We use the word "Make up your mind." And we often think "Mind–Brain," but stimuli from peripheral interaction come to our body at the peripheral as the octopus is doing and in the case of humans to inner part of our body and then to the brain. So, we have to study how we make up our minds in more detail.

Fukuda's approach leaves decision making utterly to humans. He provides the quantitative performance indicator using MD, but how the learner can decrease MD and can learn to swim is utterly left to the learner. The quantitative performance indicator in terms of MD is just a support tool for learners to how to make the next move or how to improve.

Table 9.1 Comparison of traditional engineering and strategic engineering

| | Traditional engineering | Strategic engineering |
|---|---|---|
| Base | Tactics | Strategic |
| | Problem solving | Problem definition |
| | | Goal finding |
| Perspective | Deeper and deeper | Wider and wider |
| Focus | Time | Time and space |
| Approach | Model-based | Trial and error |
| Performance indicator | Available | Need to find |
| Value | Function | Adaptability |
| | Reproducibility | Evolution |
| | Robustness | |

As IoT indicates humans and machines work together on the same team in the future. We should leave more to humans. Automation is fine. But we should remember its cost increases exponentially. If we can use humans in a more effective way, we can reduce cost tremendously and we can communicate with machines in a more harmonious way.

Table 9.1 shows comparison of traditional engineering and strategic engineering.

## References

1. E. Brown, N. Rodenberg, J. Amend, A. Mozeika, E. Steltz, M.R. Zakin, H. Lipson, H.M. Jaeger, Universal robotic gripper based on the jamming of granular material. PNAS **107**(44), 18809–18814 (2010)
2. https://en.wikipedia.org/wiki/Charles_Sanders_Peirce
3. https://plato.stanford.edu/entries/abduction/
4. https://plato.stanford.edu/entries/abduction/peirce.html
5. https://en.wikipedia.org/wiki/Walter_A._Shewhart
6. https://en.wikipedia.org/wiki/Donald_Schon
7. D.A. Schon, *The reflective practitioner: how professionals think in action* (Basic Books, New York, 1984)

The manufacturer's authorised representative in the EU is Springer Nature Customer Service Centre GmbH, Europaplatz 3, 69115 Heidelberg, Germany. If you have any concerns regarding our products, please contact ProductSafety@springernature.com

Printed and bound by CPI Group (UK) Ltd, Croydon, CR0 4YY
05/10/2026
02228152-0001